Teacher's Guide to Investigations

HOLT, RINEHART AND WINSTON

Harcourt Brace & Company

Austin • New York • Orlando • Atlanta • San Francisco • Boston • Dallas • Toronto • London

About the Teacher's Guide to Investigations

The *Modern Earth Science Teacher's Guide and Answer Keys to Investigations* combines teacher's resources for the entire three-part investigation program for *Modern Earth Science* consisting of *Small-Scale Investigations, In-Depth Investigations,* and *Long-Range Investigations.* Teacher notes and instructional strategies are provided and copies of pupil worksheets are overprinted with answers for easy access. A combined materials list is also given for pre-planning the semester.

Make sure students are familiar with the safety precautions given in their textbook and in their investigation manuals. Safety concerns are highlighted by the word **CAUTION** and the use of an appropriate safety icon. Students should review the meaning of these icons before beginning any investigation.

WARD'S Natural Science Establishment, Inc., is the official materials and equipment supplier for *Modern Earth Science.* Holt, Rinehart and Winston's unique partnership with WARD'S guarantees that:

- WARD'S lab materials and supplies are selected to fulfill the requirements of the investigations found in the *Modern Earth Science* program.

- the Small-Scale and In-Depth Investigations, which constitute the lab activities in *Modern Earth Science,* have been bench-tested by WARD'S technical staff to ensure positive experiences and successful results.

ISBN 0-03-051509-2

5   021   00   99

Contents

Teacher's Guide to
Small-Scale Investigations

About Small-Scale Investigations

As part of the overall package of investigation, the *Modern Earth Science Small-Scale Investigations* allows for easy access to the understanding of important concepts within the text through hands-on activity. Each investigation provides students with independent practice in the process of scientific inquiry. Most of the *Small-Scale Investigations* require only common household objects; others may also call for ordinary laboratory supplies. Designed for quick application, each investigation takes approximately 30 minutes to complete. Students may complete investigations at home or in the class. Answers for the analysis and conclusions questions are provided in this booklet.

M O D E R N   E A R T H   S C I E N C E

Chapter 1: Introduction to Earth Science
Small-Scale Investigation: The Big Bang Theory

According to the big bang theory, almost all galaxies are moving outward from all other galaxies. You can demonstrate the principles of this expansion with a simple model.

Materials
large (6–7 cm), uninflated round balloon; water-based felt-tip pen; string, 30 cm long; ruler

Procedure
1. Mark a pair of dots 0.5 cm apart across the middle of the uninflated balloon. Label them **A** and **B**. Mark a third dot 5.0 cm away from **B**. Label this dot **C**.
2. Blow into the balloon for 2–3 seconds. Record your elapsed time. Pinch the end of the balloon between your fingers to keep it inflated, but do not tie the neck.
3. Use the string and ruler to measure the distance between **A** and **B** and between **C** and **B**.
4. Calculate the rate of change in the distances between **A** and **B** and between **C** and **B**. To calculate the rate, subtract the original starting distance between the dots from the distance measured after inflation. Divide this number by the number of seconds you blew into the balloon.
5. With the balloon still inflated from Step 2, blow into the balloon for an additional 2–3 seconds.
6. Measure and calculate the rate of change in the distances between **A** and **B** and between **C** and **B**. To calculate the rate, use the distance measured in Step 3 as the "original" distance.

Analysis and Conclusions
1. Did the distance between **A** and **B** or between **C** and **B** show the greatest rate of change?
   The distance between **C** and **B** showed a greater rate of change.

2. Did the rate of change for either set of dots differ in Steps 4 and 6?
   Yes. The rate of change in distance for both sets of dots was larger in Step 4.

3. Suppose dots **C** and **A** represent galaxies and dot **B** represents the earth. How does the distance between the galaxies and the earth relate to the rate at which they are moving apart?
   The galaxies farther from the earth are receding at a faster rate than those

   closer to the earth.

HRW material copyrighted under notice appearing earlier in this work.

3

M O D E R N   E A R T H   S C I E N C E

Chapter 2: The Earth in Space

Small-Scale Investigation: Gravity and Orbits

A moving body tends to move in a straight line at constant speed unless some outside force acts on it. Gravity is the outside force that acts on satellites and keeps them in orbit around the earth. You can investigate the effect of gravity on a moving body with a simple model.

Materials
strip of flexible cardboard, 3×30 cm; transparent tape; sheet of white paper; a marble

Procedure
1. Form a hoop with the cardboard strip and fasten the two ends together with tape.
2. Place the hoop in the center of your paper and trace a circle along the outside edge of the hoop. Mark four points at equal distances around the circle. Number the points 1, 2, 3, and 4.
3. Place a marble inside the cardboard hoop. Slowly swirl the hoop clockwise until the marble rolls smoothly around the inside edge of the hoop. Stop swirling the hoop as the marble approaches point l; then quickly lift the hoop, allowing the marble to escape. You may have to practice this step several times to get the marble released at the right time.
4. Observe and record the path of the marble as it exits the hoop.
5. Repeat Steps 3 and 4, stopping the hoop and releasing the marble at points 2, 3, and 4.

Analysis and Conclusions
1. What path does the marble take when the hoop is removed? Is the pattern of the path the same for all four exit points?

   a straight path; yes

2. In what direction does the hoop push the marble? What force does the hoop represent?

   toward the center of the circle; gravity

3. Compare the motion of the marble with that of a satellite around the earth. How are they alike? How are they different?

   Just as the marble was held in a circular path by the hoop, a satellite is held in

   orbit by the earth's gravitational force. However, the earth's gravity, unlike the

   hoop, is an invisible force.

M O D E R N   E A R T H   S C I E N C E

Chapter 3: Models of the Earth

Small-Scale Investigation: Topographic Maps

Contour lines show elevation and landforms on topographic maps. You can use contour lines to make a topographic map of a model mountain.

Materials

modeling clay (1–2 lb.); paper clip; large waterproof container, at least 8 cm deep; water; ruler; pencil; adhesive or masking tape

Procedure

1. Make a mountain 6–8 cm high out of modeling clay. Work on a flat surface and smooth out the mountain's shape, making the mountain slightly steeper on one side.
2. Run a paper clip down one side of the model to form a valley several millimeters wide.
3. Place the model in the center of the container. Tape the ruler upright in the container with one end resting on the bottom of the container. Make sure the container is resting on a level surface.
4. Add water to the container to a depth of 1 cm, using the ruler as a guide. With a sharp pencil, trace around the model and inscribe the clay along the waterline.
5. Raise the water level 1 cm at a time until you reach the top of the model. Each time you add water to the container, inscribe another contour line in the clay where the waterline meets the model.
6. When finished, carefully drain the water and remove the model from the container.

Analysis and Conclusions

1. What is the contour interval on your model?

   1 cm

2. Observe your model from directly above. Try to duplicate the size and spacing of the contour lines on a sheet of paper.

   Drawings will vary. Typical drawings will show contour lines that are irregularly

   spaced and are arranged in a somewhat circular pattern.

3. Compare the contour lines on a steep slope with those on a gentle slope. How do they differ?

   Contour lines on a steep slope are closer together than those on a gentle slope.

4. How is a valley represented on your topographic map?

   It is represented with contour lines that form a V. The V points toward the higher

   end of the valley.

M O D E R N   E A R T H   S C I E N C E

Chapter 4: Plate Tectonics

Small-Scale Investigation:
Lithospheric Plate Boundaries

The movement of lithospheric plates has created many of the earth's topographical features. You can demonstrate the results of plate movement by using clay models of lithospheric plates.

Materials
ruler, paper, scissors, rolling pin or rod, modeling clay (2–3 lb.), plastic knife, lab apron

Procedure
1. Draw two 10 × 20 cm rectangles on your paper, and cut them out.
2. Use a rolling pin to flatten out two pieces of clay until they are about 1 cm thick. Cut each piece into a 10 × 20 cm rectangle. Place a paper rectangle on each piece of clay.
3. Place the two clay models side by side on a flat surface, paper side down. Place your hands directly on top of each piece, and slowly push the models together until the edges begin to buckle and rise off the surface of the table.
4. Turn the clay models around so that the unbuckled edges are touching. If these edges have been slightly deformed during Step 3, smooth them out before proceeding.
5. Place one hand on each clay model. Apply only slight pressure toward the seam. Slide one clay model forward and the other model backward about 7 cm.
6. Repeat Step 5 three more times, alternating the direction in which you push each model.

Analysis and Conclusions
1. What type of plate boundary are you demonstrating with the model in Step 3?
   a convergent boundary
   _____

   _____

   _____

   _____

2. What type of plate boundary are you demonstrating in Steps 5 and 6?
   a transform fault boundary
   _____

   _____

   _____

3. How does the appearance of the facing edges of the models in the two processes compare? How do you think these processes might affect the appearance of the earth's surface?
   With head-on motion, the edges buckle and rise; with sliding motion, the edges

   become rough and jagged. Head-on collisions cause the earth's crust to crumple

   and mountains to form. Sliding causes the earth's crust to fracture.

   _____

   _____

   _____

6

M O D E R N   E A R T H   S C I E N C E

Chapter 5: Deformation of the Crust
Small-Scale Investigation: Folds and Fractures

You can use some common objects to demonstrate the factors that govern the ways rock responds to stress from plate movements.

Materials
safety goggles; soft wood dowel, 2 mm × 15 cm; 2 books; plastic play putty

Procedure
1. **Put on the safety goggles.** Lay the dowel on a table. Place a book at each end of the dowel.
2. Place one hand on each book and gently and slowly slide the books against the ends of the dowel until the dowel bends slightly.
3. Move the books back to their original position. Observe and record what happens to the dowel as the books are moved.
4. Again move the books toward each other with the dowel between them. This time move the books quickly and forcefully. Record what happens to the dowel. Remove the safety goggles.
5. Roll a piece of play putty into a cylinder about 15 cm long and about the same diameter as the dowel.
6. Repeat Steps 1 through 3, using the putty in place of the dowel.
7. Re-form the putty into a cylinder. Grasp one end of the cylinder in each hand and pull quickly and sharply on both ends of the putty.

Analysis and Conclusions
1. Compare the responses of the dowel and the putty in Step 3. What two responses of rock to stress are represented?

   The dowel bends but regains its original shape when the books are removed. The

   putty bends, but it does not regain its original shape. The dowel represents rock

   that is deformed temporarily by stress but regains its shape when the stress is

   removed. The putty represents rock that permanently folds as a result of stress.

   _____

2. Compare the response of the dowel in Step 4 with the response of the putty in Step 7.

   Both the dowel and the putty fracture.

   _____

   _____

3. What two factors influence the way the items respond to stress in this investigation? How do these factors influence the way rock responds to stress? Explain your answer.

   The two influential factors are the physical properties of the material and the

   type of stress. Rocks deep within the earth's crust are made more pliable by the

   high temperatures. Cooler temperatures at the earth's surface make rock brittle.

   Rocks deformed by slow, gentle stress may regain their original shape when the

   stress is removed. Quick, strong stress usually fractures rock.

   _____

M O D E R N   E A R T H   S C I E N C E

Chapter 6: Earthquakes

Small-Scale Investigation: Seismographic Record

A seismograph records the energy release of an earthquake. You can observe energy release by making a model seismograph.

Materials
shoe box, large plastic bag, sand (15–20 lb.), felt-tip pen, rubber band or masking tape, pad of paper or clipboard, four objects with various masses (max. 1–2 kg), balance, metric ruler, newspaper

Procedure
1. Line the shoe box with the plastic bag, and fill the box with sand. Put on the lid.
2. Mark an X near the center of the lid.
3. Fasten the felt-tip pen to the lid of the box with a tight rubber band or masking tape so that it just extends beyond the edge of the box.
4. Measure and record each test object's mass.
5. Have a partner hold the pad of paper so that it just touches the pen.
6. Hold your first object directly over the X at a height of about 30 cm. As your partner slowly moves the paper horizontally past the pen, drop the object on the X.
7. Label the resulting line with the mass of the object dropped and the material in the box.
8. Repeat Steps 5–7 with three more objects of different masses. Move the starting position of the paper up about 2 cm each time so you will end up with four lines at 2 cm intervals.
9. Replace about 2/3 of the sand with crumpled newspaper. Reassemble the box and repeat Steps 5–8.

Analysis and Conclusions
1. What do the lines on the paper represent?

   energy release
   _____

   _____

2. What do the sand and newspaper represent?

   different types of crustal material
   _____

   _____

3. Compare the lines made in Steps 6–8 with those made in Step 9. Explain any differences.

   For a given mass, the box filled with sand shows less displacement than the

   box containing sand and newspaper. The box filled with sand is more rigid and

   thus vibrates less.
   _____

4. What can you infer about the relationship between seismographic records and the energy released by an earthquake?

   Increased vibration produces more-erratic lines, which suggests greater energy

   release.
   _____

   _____

   _____

M O D E R N   E A R T H   S C I E N C E

Chapter 7: Volcanoes

Small-Scale Investigation: Volcanic Cones

A cinder cone is formed by material thrown out during an explosive volcanic eruption. A shield cone is made up of layers of lava that have flowed out during a quieter eruption. You can demonstrate that the very different shapes of these cones are the result of the different materials from which they are made.

Materials
plaster of Paris; measuring cup; water; mixing spoon; metric ruler; 2 paper plates; protractor; dry cereal or potato flakes, 8 oz.; graduated cylinder, 100 mL

Procedure
1. Pour 1/2 cup (about 4 oz.) of plaster of Paris into the measuring cup. Gently tap the cup so that the plaster settles to the 1/2 cup level.
2. Use the graduated cylinder to measure out 60 mL of water, and add it to the dry plaster in the measuring cup. Use a mixing spoon to blend the mixture until it is smooth and uniform.
3. Hold the measuring cup about 2 cm over a paper plate. Pour the contents slowly and steadily onto the center of the plate.
4. When the cone has hardened (15–20 min.), remove it from the plate. Measure the average slope angle with a protractor.
5. Pour dry cereal or potato flakes slowly onto the center of a clean paper plate until the mound is approximately 5 cm high.
6. Without disturbing the mound, measure its slope with a protractor.

Analysis and Conclusions
1. Which cone that you formed represents a cinder cone? a shield cone? How do the angles formed by these cones compare?

   The cone made of cereal represents a cinder cone. The cone made of plaster of

   Paris represents a shield cone. The cone made of dry cereal or potato flakes will

   have a steeper slope than the cone made of plaster of Paris.

2. How would the slope have been affected if the cereal was rounder? thicker?

   Answers will vary. A typical answer would be that a cone made of rounder cereal

   would not hold together at all. The rounder pieces of cereal would roll away and

   not hold together. Thicker cereal would probably scatter rather than form a cone.

3. Suppose you formed a cone by pouring alternating layers of wet plaster of Paris and dry cereal. How would the shape and overall size of this cone differ from the two cones you have already formed? Which type of volcano would such a cone look like?

   The layered cone would have a broader base than a cinder cone and would be

   higher than a shield cone. The cone would resemble a composite cone.

M O D E R N   E A R T H   S C I E N C E

Small-Scale Investigation:
Electrically Charged Objects

Electrically charged objects behave in predictable ways. Objects with unlike charges attract each other; objects with like charges repel each other. Objects with a neutral charge neither attract nor repel each other. You can demonstrate this behavior with several common objects.

Materials
balloon, thread, ruler, rubber or plastic comb, drinking glass, wool cloth, plastic bag, books

Procedure
1. Inflate the balloon and hang it from the ruler by a piece of thread. Hold the comb next to the balloon and observe.
2. Repeat Step 1 using the glass and not the comb.
3. Rub the balloon and the comb with the wool cloth. Hold the comb next to the balloon.
4. Rub the glass with the plastic bag. Hold the glass next to the balloon.

Analysis and Conclusions
1. Describe the behavior of the balloon, comb, and glass in Steps 1 and 2. What can you conclude about their electrical charges?

   Nothing happens. The objects must have neutral charges because objects with

   neutral charges do not attract or repel each other.

2. Describe the behavior of the balloon and comb in Step 3. What can you conclude about their electrical charges?

   The balloon and comb repel each other. Both objects must have picked up the

   same charge when they were rubbed with the wool cloth because objects with

   like charges repel each other.

3. Describe the behavior of the balloon and glass in Step 4. What can you conclude about their electrical charges?

   The balloon and glass attract each other. The objects must have unlike

   charges because objects with unlike charges attract each other.

4. Rubbing the glass with the plastic bag produces a positive electrical charge on the glass. What is the charge of the balloon in Steps 3 and 4? What is the charge of the comb in Step 3?

   Because objects with unlike charges attract each other and the glass has a

   positive charge, the balloon must have a negative charge. Because objects with

   like charges repel each other, the comb must also have a negative charge.

5. Two ions, such as positive sodium and negative chloride, may form a chemical bond. What can you infer about the charges of the ions?

   Because a chemical bond is the result of electrical attraction, the ions must have

   opposite charges.

M O D E R N   E A R T H   S C I E N C E

Chapter 9: Minerals of the Earth's Crust
Small-Scale Investigation: Mineral Identification

Most minerals can be identified by two or three physical properties. You can use some basic tests of mineral properties and a table of mineral characteristics to identify an unknown mineral.

Materials
mineral specimen, streak plate, Mohs hardness scale on page 165 of your textbook, copper penny, steel file, glass plate, Guide to Common Minerals on pages 666–667 of your textbook.

Procedure
1. Use the data table below to record your observations.
2. Find a mineral specimen that seems to be the same color and makeup throughout. Record the color of your specimen.
3. If your specimen shines like polished metal, consider its luster metallic; otherwise, nonmetallic. Record your observations.
4. Rub your specimen across the streak plate. Record the color of the streak.
5. Study Mohs hardness scale. Use your fingernail, a penny, a glass plate, and finally a steel file to scratch your specimen. Match the item that scratched your specimen with those listed beside the scale. Record your results.
6. Compare your results with the Guide to Common Minerals. List all those minerals that fit the description of your specimen.
7. Refer to the guide again and make a list of five minerals that clearly do not fit the description of your specimen.

Analysis and Conclusions
1. Which tests were the most useful in selecting the minerals that fit the description of your specimen? Tentatively identify the type of mineral you have.

   Answers will vary. For minerals with a metallic luster, the color of the streak is

   very useful. For minerals with a nonmetallic luster, the hardness is very useful.

   Minerals identified should have the same properties as the student's specimen.

2. Suggest other tests that would positively identify your specimen.

   Answers will vary. Typical answers may include crystal form, cleavage, density,

   and reaction to dilute hydrochloric acid.

Color	
Luster	
Streak	
Hardness	

M O D E R N   E A R T H   S C I E N C E

Chapter 10: Rocks

Small-Scale Investigation: Crystal Formation

The rate of cooling affects the size of the crystals of minerals found in igneous rock. You can demonstrate this relationship by cooling crystals of Epsom salts at three different rates.

Materials

3 glasses or glass jars, water, ice cubes, 3 large test tubes, small pan, spoon or stirring rod, measuring cup, Epsom salts, stove or hot plate, tongs or test-tube clamp, cork, clock or watch

Procedure

1. In a small saucepan, mix 120 mL of Epsom salts in 120 mL of water. Heat over low heat. Do not let the mixture boil. Stir until no more crystals will dissolve.
2. Add the following until each glass is 2/3 full: glass 1—water and ice cubes; glass 2—water at room temperature; glass 3—hot tap water
3. Carefully pour equal amounts of the Epsom salts mixture into the 3 test tubes. Use the tongs to steady the test tubes as you pour. Drop a few crystals of Epsom salt into each test tube and gently shake. Place one test tube into each glass.
4. Observe what happens to each of the solutions as they cool at different rates over the next 10–15 minutes. Let the glasses sit overnight, and examine the solutions again after 24 hours.

Analysis and Conclusions

1. In which test tube are the crystals the largest? the smallest?

   The crystals are the largest in the test tube in the hot water and the smallest

   in the test tube in the cold water.

2. How does the rate of cooling affect the size of the crystals formed? Explain your answer.

   As the cooling rate increases, the size of the crystals formed decreases. The

   warmest solution had the slowest cooling rate and produced the largest crystals.

3. How would you change the procedure to obtain even larger crystals of Epsom salts? Why?

   Place a test tube of the salt solution in water that is close to the boiling point.

   The hotter the water, the longer it takes to cool. The longer cooling time allows

   larger crystals to form.

4. Some igneous rocks that are thrown out of a volcano contain large crystals surrounded by very small ones. Based on your observations in this activity, explain why the crystals are different sizes.

   While inside the volcano, the rock was cooling at a slow rate, so large crystals

   were formed. As the rock was hurled through the air, it cooled at a more rapid

   rate, thus forming the smaller crystals in the outer part of the rock.

M O D E R N   E A R T H   S C I E N C E

Chapter 11: Resources and Energy

Small-Scale Investigation: Solar Collector

You can demonstrate the principles of solar collection with a simple model.

Materials

small shallow pan, sheet of black plastic, thermometer, adhesive tape, water at room temperature, plastic wrap, rubber bands, clock or watch

Procedure

1. Line the inside of the pan with black plastic. Tape the thermometer upright to the inside of the pan. Fill the pan with water at room temperature just enough to cover the end of the thermometer. Fasten plastic wrap over the pan with a rubber band. Be sure you can read the thermometer.
2. Place the pan in a sunny area. Record the temperature every 5 minutes until it stops rising. Discard the water.
3. Repeat the procedure in Steps 1 and 2, but do not cover the pan with plastic wrap.
4. Repeat the procedure in Steps 1 and 2, but this time do not line the pan with black plastic.
5. Repeat the process once again, but do not line the pan with black plastic and do not cover the pan with plastic wrap.
6. Calculate the rate of temperature change for each trial. To do so, subtract the beginning temperature from the ending temperature. Divide this number by the number of minutes it took until the temperature stopped rising.

Analysis and Conclusions

1. What are the variables in this investigation? Which model had the greatest rate of temperature change? the least?

   the black plastic, the plastic wrap; the pan lined with black plastic and covered

   with plastic wrap; the pan not lined with black plastic and not covered with

   plastic wrap

2. Which variable that you tested has the most significant effect on temperature change?

   The black plastic had the most significant effect. It absorbed energy from the

   sun, and the plastic wrap helped store the energy in the pan so the energy could

   heat the water.

3. What materials would you use to make an efficient solar collector of your own design?

   Answers will vary. Typical answers will emphasize the importance of using

   materials that absorb sunlight like the black plastic in this investigation.

MODERN EARTH SCIENCE

Chapter 12: Weathering and Erosion
Small-Scale Investigation: Mechanical Weathering

You can demonstrate the effects of mechanical weathering by abrasion by placing rocks in a container of water and shaking the container.

Materials
silicate rock chips; hand lens; 2 plastic containers (16 oz. or larger), one with a tight-fitting lid; water; strainer; clear glass jar or small beaker

Procedure
1. Examine the rocks with a hand lens, noting the shape and surface texture.
2. Fill the plastic container with the tight-fitting lid about half full of rocks. Add water to barely cover the rocks.
3. Tighten the lid, and shake the container 100 times.
4. Hold the strainer over the other container. Pour the water and rocks into the strainer.
5. Run your finger around the inside of the empty container. Write down what you feel.
6. Use the hand lens to observe the rocks.
7. Pour the water into the glass jar, and examine the water with the hand lens.
8. Put the rocks and water back into the container with the lid. Repeat Steps 3–7.
9. Repeat Step 8 two more times.

Analysis and Conclusions
1. Has the amount and particle size of the residue left in the container changed during the course of the investigation? Explain your answer.

   Answers will vary. A typical answer would be that the amount of residue

   increased and the particle size decreased with the number of shakes.

   _____

   _____

2. How has the appearance of the rocks changed? How has the appearance of the water changed?

   Answers will vary. A typical answer would be that the rocks show rounding and

   breaking. The water should also appear dirtier with each trial as small chips and

   particles of rock accumulate there.

   _____

   _____

3. If the water from a small stream ran over a ledge of rock into a pool below, what would you expect to find at the bottom of the pool?

   Answers will vary. A typical answer would be that the pond bottom would

   contain rock sediment because of mechanical weathering of the rock ledge.

   _____

   _____

M O D E R N   E A R T H   S C I E N C E

Chapter 13: Water and Erosion

Small-Scale Investigation: Soil Erosion

The uncontrolled runoff of surface water is a major cause of soil erosion. You can demonstrate soil erosion by making a model of a hillside.

Materials

rectangular pan about 23 × 33 cm, fine sand, sink lined with paper towels, brick or other support, water faucet, additional pan or container to catch water and sand, small ruler or straightedge, clock or watch

Procedure

1. Fill the pan about half full with moist sand.
2. Place the pan in the sink so that the end resting on the brick is under the water faucet.
3. Place the second pan or container so it catches sand and water that flow out of the first pan.
4. Slowly open the faucet until a gentle trickle of water falls onto the sand in the raised end of the pan. Let the water run for 15–20 seconds.
5. Turn off the water and draw the pattern of water flow over the sand.
6. Press the sand back into place and carefully smooth the surface with a small ruler or straightedge. Repeat Step 4, but this time increase the rate of water flow slightly, but not enough to splash the sand. Draw a pattern of water flow as before.
7. Repeat Steps 4 and 5 two more times with different rates of water flow. For each trial, observe and draw the pattern of water flow.

Analysis and Conclusions

1. Compare the rate of erosion with the rate of water flow.

   The rate of erosion increases as the rate of water flow increases.

2. Compare the patterns of erosion with the rate of water flow.

   Answers may vary. As the rate of water flow increases, the effects of erosion

   become more severe.

3. How does the rate of water flow affect gullies?

   Gullies become deeper and wider as the rate of water flow increases.

4. Without changing the rate of water flow, how could the rate and effects of erosion be reduced on an actual hillside?

   Answers will vary. Typical answers will include various soil conservation

   methods, such as the use of cover plants.

M O D E R N   E A R T H   S C I E N C E

Chapter 14: Groundwater and Erosion

Small-Scale Investigation: Permeability

You can demonstrate permeability and calculate the rate of drainage with a simple model.

Materials

sharpened pencil, 3 large paper or plastic cups (9 oz.), cheesecloth, rubber bands, ruler, sand, soil, gravel, 3 small thread spools or other supports, saucer or tray, measuring cup or graduated cylinder, water, clock or watch

Procedure

1. With the pencil point, make seven tiny holes in the bottom of each cup. Cover the holes with cheesecloth secured tightly by a rubber band.
2. Mark a line 2 cm from the top of each cup. Place the first cup on the spools in a saucer, and fill the cup to the line with sand.
3. Pour 120 mL of water into the cup. Time and record how long it takes for the water to drain through the cup.
4. Pour the water from the saucer into the measuring cup or graduated cylinder. Record the amount of water.
5. Repeat Steps 3 and 4 with the two other cups—one with soil, the other with gravel.
6. Calculate the rate of drainage for each cup by dividing the amount of water that drained by the time it took the water to drain.
7. Calculate the percentage of water retained in each cup by subtracting the amount of water that drained into the saucer from the original 120 mL. Divide the difference by 120.

Analysis and Conclusions

1. Which cup had the highest drainage rate?

   A typical answer would be either gravel or sand.

   _____

   _____

   _____

2. Which cup retained the least water?

   A typical answer would be either gravel or sand.

   _____

   _____

   _____

3. Consider the sample with the highest drainage rate and lowest percentage of water retained to be the most permeable. Which sample is the most permeable? the least permeable?

   A typical answer would be that either gravel or sand is the most permeable and

   soil is the least permeable.

   _____

   _____

M O D E R N   E A R T H   S C I E N C E

Chapter 15: Glaciers and Erosion
Small-Scale Investigation: Glacial Erosion

Many of the earth's natural features have been shaped by the movement of glaciers. You can demonstrate the effects of glacial erosion with a simple model.

Materials
plastic container (about $15 \times 10 \times 5$ cm); mixture of sand, gravel, and small rocks; water; freezer; modeling clay (1–2 lb.); hand towel; sand (1–2 lb.); flat box (about 30×20 cm); soft wood board (about 30 cm long); rolling pin or large dowel

Procedure
1. Put the sand, gravel, and rock mixture in the bottom of the plastic container. Fill the container with water to a depth of about 4 cm. Freeze the container until the water is solid. Remove the ice block from the container.
2. Use a rolling pin or large dowel to flatten the modeling clay into a rectangle about $20 \times 10 \times 1$ cm. Grasp the ice block firmly with the hand towel. Place the block with the gravel-and-rock side down at one end of the clay. Press down on the ice block and move it along the length of the flat clay surface. Sketch the pattern made in the clay by the ice block.
3. Next, press damp sand into the bottom of a shallow rectangular box. As in Step 2, move the ice block along the surface of the sand while pressing down lightly.
4. Repeat the process outlined in Steps 2 and 3 using a soft wooden board in place of the clay and sand.

Analysis and Conclusions
1. Describe the effects of moving the ice block over the clay, sand, and wood.
   The movement of the ice block made gouges in the surface of the clay, pushed

   clay into the sides of the ice block, and pushed clay up in front of the ice block.

   A similar but more pronounced pattern resulted when the ice block moved

   through the sand. The ice block made scratches and gouge marks in the surface

   of the wood.

2. Did any material from any surface become mixed with material from the ice block? Did the ice deposit material on any surface?
   Answers will vary. A typical answer would be that small amounts of the surface

   material became mixed with the ice and small amounts of the material in the ice

   were deposited on the sand and clay.

3. What glacial land features are represented by the features of your clay model? Your sand model? Your wood model?
   Answers will vary. Typical answers will include moraines, erratics, and drumlins

   for both the clay and sand models, and scratches or gouges for the wood model.

4. Based on your observations, predict the results of glacial erosion on rock.
   The glacier would not displace the rock as it does softer materials. It would make

   gouges and scratch marks in the rock and would leave the rock with more

   rounded edges.

17

M O D E R N   E A R T H   S C I E N C E

Chapter 16: Erosion by Wind and Waves
Small-Scale Investigation: Dune Migration

You can demonstrate how dunes migrate by making a model dune and simulating natural events.

Materials
wax pencil or marker; ruler; shallow box or tray; paper bag, large enough to hold one end of the box; very fine, dry sand; safety goggles; filter mask; hair dryer; clock or watch

Procedure
1. Use a marker and ruler to mark the outside of the box at 5-cm intervals.
2. Place the box halfway inside the paper bag so the bag will catch any blowing sand.
3. Fill the box about half full of fine, dry sand. Make a dune in the sand about 10–15 cm from the open end of the box. Look at the side of the box and record to the nearest centimeter the location of the peak of the dune.

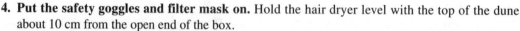

4. **Put the safety goggles and filter mask on.** Hold the hair dryer level with the top of the dune about 10 cm from the open end of the box.
5. Turn the hair dryer to low speed for 1 minute. Identify and record the new location of the peak of the dune to the nearest centimeter.
6. Repeat Step 5 three times, first running the hair dryer for 2 minutes, then 3 minutes, and then 5 minutes. After each trial, record the location of the peak of the dune.
7. Flatten the sand. Place a barrier such as a rock in the sand. Position the hair dryer as in Step 4. Run the dryer for 3 minutes.

Analysis and Conclusions
1. In what direction did the dune migrate?

   The dune migrates away from the wind source.

2. How far overall did the dune migrate?

   Answers will vary. A typical answer would be about 5 to 10 cm.

3. What was the average distance the dune migrated per minute?

   Answers will vary. A typical answer would be about 0.5 to 1 cm.

4. In Step 7, where does the dune form? What steps might be taken to slow down the process of dune migration? Explain your answer.

   The dune forms on the downwind side of the barrier.

   Answers will vary. A typical answer would be that a dune will migrate until it

   reaches a barrier, so dune migration could be slowed down by planting grasses

   or shrubs or by constructing windbreaks.

M O D E R N   E A R T H   S C I E N C E

Chapter 17: The Rock Record

Small-Scale Investigation: Radioactive Decay

Radioactive elements decay at a constant, measurable rate. The time it takes for half of any given amount of an original element to change into a new isotope or element is called a *half-life*. You can demonstrate the principle of radioactive decay with a simple model.

Materials
clock or watch with a second hand; sheet of ruled notebook paper, about 28×22 cm; scissors

Procedure
1. Record the time.
2. Wait 20 seconds, then carefully cut the sheet of paper in half. Select one piece, and set the other piece aside.
3. Wait 20 seconds, then cut the selected piece of paper in half. Select one piece, and set the other piece aside.
4. Repeat Step 3 until nine 20-second intervals have elapsed.

Analysis and Conclusions
1. In terms of radioactive decay, what does the whole piece of paper used in this investigation represent?

   a radioactive parent element
   _____
   _____

2. What do the pieces of paper that you set aside in each step represent?

   the nonradioactive daughter element that has formed as a result of radioactive
   decay
   _____

3. What is the half-life of your "element"?

   20 seconds
   _____
   _____

4. How much of your paper "element" was left after the first three intervals? after six intervals? after nine intervals? Express your answer as percentages.
   12.5%; 1.56%; 0.195%
   _____

5. What two factors in your model must remain constant for your model to be accurate? Explain your answer.
   The interval between the cuts and the proportions of each piece of paper cut
   must remain constant. Radioactive elements decay at a constant rate. In order
   for the model to be accurate, these factors must not change.
   _____
   _____

MODERN EARTH SCIENCE

Chapter 18: A View of the Earth's Past

Small-Scale Investigation: Geologic Time Scale

Geologic time spans some 4.6 billion years. Using a scale that represents time as distance, you can compare the length of eras and periods.

Materials

geologic time scale on pages 354–355 of your textbook; adding-machine paper, 5 m long; ruler; meter stick; 6 colored pencils

Procedure

1. The table below lists the major eras of the geologic time scale and their lengths in years.
2. Complete the table, using the scale 1 cm equals 10 million years.
3. Lay the adding-machine paper flat on a hard surface. Use the meter stick and a pencil to mark off the beginning and end of Precambrian time according to the time scale you calculated. Do the same for the other eras. Label the eras and color each era a different color.
4. Study the geologic time scale on pages 354–355 of your textbook. Use the scale 1 cm equals 10 million years and calculate the scale length for each period listed.
5. Mark the boundaries of each period by drawing a line across the width of the adding-machine paper. Label the periods on your scale.
6. Write in the major kinds of organisms that lived during each unit of time.

Analysis and Conclusions

1. Does the scale you made measure relative time, absolute time, or both?

   The scale shows both relative and absolute time. The scale shows the duration

   of eras and periods relative to each other. However, because estimated dates are

   known and written on the scale, the scale is also a measure of absolute time.

2. When did humans first appear? What is the scale length from that period to the present?

   during the Quaternary Period; about 0.25 cm

3. Add the lengths of the Paleozoic, Mesozoic, and Cenozoic eras. What percentage of the total geologic time scale do these eras combined represent? What percentage of the total time scale does Precambrian time represent?

   The combined duration of the three eras is 570 million years and represents only

   about 12 percent of the total geologic time scale. Precambrian time represents

   the other 88 percent.

Geologic Time Scale		
Era	Length of time (years)	Scale length (cm)
Precambrian	4,030,000,000	403
Paleozoic	325,000,000	32.5
Mesozoic	180,000,000	18
Cenozoic	65,000,000	6.5

M O D E R N   E A R T H   S C I E N C E

Chapter 19: The History of the Continents

Small-Scale Investigation: Plate Tectonics Theory

Although questions still remain about the movement of lithospheric plates, most scientists think that plate movement is caused by convection currents in the asthenosphere. You can demonstrate this theory by making a model of the asthenosphere and lithospheric plates.

Materials

water; shallow rectangular pan, about 23 cm × 33 cm; stove or 2 hot plates; dark-colored food coloring; 8 pieces of shirt-packaging cardboard, about 1 cm² each; marker; metric ruler

Procedure

1. Fill the pan with water until it is 3 cm from the top, and place it on the stove or hot plates.
2. Heat the water for 30 seconds over very low heat. Add a few drops of food coloring to the center of the pan. Observe and record what happens to the food coloring. Turn off the heat.
3. Label the cardboard pieces 1 through 8. Carefully place pieces 1 through 4 as close together as possible in the center of the pan. Place one of the four remaining cardboard pieces in each corner of the pan.
4. Turn on the heat. Sketch the pattern of movement for each cardboard piece. Turn off the heat. **Caution: Use a pot holder when handling heated objects.**

Analysis and Conclusions

1. In the plate tectonics model, what do the water in the pan, the cardboard pieces, and the stove represent?

   asthenosphere; lithospheric plates; the earth's core

2. In Step 2, what happened to the food coloring in the water? Explain your answer.

   The food coloring moved through the water in a circular pattern. This pattern

   was created by convection currents caused by heating the water in the pan.

3. Describe what happened to the cardboard pieces as the water was reheated in Step 4.

   At first, the cardboard pieces in the center of the pan moved apart from one

   another in a circular pattern. The cardboard pieces in the corners of the pan did

   not move unless touched by one of the other pieces. Eventually, all of the

   pieces moved—some in a clockwise circle, others in a counterclockwise circle.

4. In what way is the model you made an inaccurate representation of the movement of lithospheric plates?

   Answers will vary. A typical answer would be that the heat source in the model

   was closer to some of the cardboard pieces, which might have caused some

   pieces to begin moving before others. Heat from the earth's core is more evenly

   distributed than the heat in the model.

M O D E R N   E A R T H   S C I E N C E

Chapter 20: The Ocean Basins
Small-Scale Investigation: Sonar

You can demonstrate the principle of sonar by calculating the rate at which pulses travel along a spring to determine distance.

Materials
toy spring coil or flexible spring, heavy string, meter stick, stopwatch, adhesive tape

Procedure
1. Use heavy string to tie one end of the spring securely to a doorknob. Pull the free end of the spring taut and parallel to the floor. Keep the tension of the spring constant throughout the investigation.
2. Use tape to mark the floor directly beneath the hand holding the spring taut. Measure and record the distance from your hand to the doorknob.
3. Note the time. Hold the spring taut and hit the spring sharply with the side of your free hand. Strike the spring as close as possible to your hand holding the spring.
4. Check the time again to see how long it takes for the pulse to travel to the doorknob and back to your hand.
5. Repeat Steps 2–4 three times. Each time hold the spring 60 cm closer to the doorknob than in the previous trial, keeping the tension constant by gathering coils into your hands as necessary.
6. Calculate the rate of travel for each trial by multiplying the length of the spring between your hand and the doorknob by 2. Then divide by the number of seconds it took for the pulse to travel to the doorknob and back.

Analysis and Conclusions
1. Did the rate of travel of the pulse change during the course of the investigation?
   No. The rate remains the same for each trial.

2. If a pulse took 3 seconds to travel to the doorknob and back to your hand, what is the distance from the doorknob to your hand?
   Answers will vary. A typical answer would be that by knowing the rate of travel of the pulse in the spring, you can calculate distance.

3. How is the apparatus you used like sonar? How is it different? Explain your answer.
   Answers will vary. A typical answer would be that the pulse created in the spring models the sound waves used in sonar; however, unlike the spring apparatus, sonar can determine the position of objects that cannot be seen and cannot be reached.

M O D E R N E A R T H S C I E N C E

Chapter 21: Ocean Water

Small-Scale Investigation: Density Factors

Temperature and salinity are two variables that influence the density of water. You can test these two variables using a simple straw float.

Materials

glass jar (1-L), distilled water, plastic soda straw, modeling clay, scissors, table salt, freezer, thermometer, metric ruler, marker, measuring cup

Procedure

1. Fill the jar with about 1 L of distilled water at room temperature.
2. Mark the length of the straw at 1-cm intervals. Fill 5 cm of one end of the straw with modeling clay.
3. Float the straw upright in the jar. If the straw does not float upright, cut off the open end at 1-cm intervals until it floats upright.
4. As the straw floats in the jar, use the 1-cm markings to estimate the length of the straw that remains below the water surface. Record your observations.
5. Put the jar in the freezer until the temperature of its contents drops to about 8°C. Remove the jar from the freezer and repeat Step 4. Record your observations.
6. Wait until the water returns to room temperature. Add about 1/8 cup of salt to the water. Stir the water until the salt is completely dissolved. Repeat Step 4. Record your observations.

Analysis and Conclusions

1. In which trial was the water most dense? the least dense? Explain your answers.

   Answers will vary. Typical answers will include salt water, cold water, and

   distilled water, ranked from most dense to least dense. As the density

   decreases, the straw will sink deeper below the water's surface.

2. Based on your observations, how might you make the distilled water in Step 5 less dense?

   Answers will vary. A typical answer would be to warm the water.

3. What effect would heating the water have on density?

   It would decrease the water's density.

4. Based on your observations, where would you expect the water in the ocean to be the least dense? the most dense?

   Answers will vary. A typical answer would be that the water will be least dense

   in areas where the water has a low salinity or is very warm, or where there is an

   infusion of fresh water. The water will be most dense in areas where the water

   has a high salinity or is very cold.

M O D E R N   E A R T H   S C I E N C E

Chapter 22: Movements of the Ocean

Small-Scale Investigation: Waves

As a wave passes through water, the form of the wave moves across the surface. However, a wave does not carry water along with its motion. You can demonstrate this principle of wave movement with some common objects.

Materials

sink or rectangular pan (at least 40 cm $\times$ 30 cm $\times$ 10 cm), water, ruler, cork stopper, adhesive tape, teaspoon

Procedure

1. Fill the pan with water to a depth of 7 cm. Float the cork near the center of the pan. On each side of the pan, mark the location of the cork with a small piece of tape.
2. Hold the spoon in the water, and slowly and carefully move it up and down in the water.
3. Continue to make waves in a slow, regular pattern. Observe the movement of the cork. Sketch what you have observed.
4. Remove the cork from the pan. Again use the spoon to make waves. This time, pay careful attention to creating a strong, steady series of waves. Remove the spoon from the pan. Observe what happens when the waves reach the edges of the pan. Write down or sketch what you observed.

Analysis and Conclusions

1. Describe the motion of the cork when a wave passes.

   Answers will vary. A typical answer would be that the cork seems to move

   slightly forward at first and then slightly backward. The motions forward and

   backward are equal.

   _____

2. How does the cork move relative to the tape on the side of the pan? Explain your answer.

   The cork stays almost even with the tape on the pan. The general motion of the

   cork is to bob up and down and slightly forward, then back. The cork does not

   show any significant net movement horizontally in the direction of the wave.

   _____

3. When a wave breaks on the shore of a beach, it is clear that the water is carried forward with the form of the wave. Based on your observations in Step 3, does this contradict your model? Explain.

   No. The waves created in the pan resemble waves in deep water rather than

   waves breaking on the shore of a beach. Because the pan is the same depth

   throughout, the water moves back and forth in the pan instead of being carried

   forward by passing waves. If the pan was shallower at one end, then the waves

   would break as they moved into the shallow water and would carry water

   forward.

   _____

   _____

M O D E R N   E A R T H   S C I E N C E

Chapter 23: The Atmosphere

Small-Scale Investigation: Barometric Pressure

A barometer measures changes in atmospheric pressure. You can construct a simple aneroid barometer with some common objects.

Materials

plastic wrap; coffee can with a diameter of about 10 cm, open at one end; rubber band; drinking straw, about 10 cm long; masking or adhesive tape; cardboard, 10 cm wide and at least 8 cm taller than the can; metric ruler

Procedure

1. Refer to the photo on page 458 of your textbook when making your barometer.
2. Secure plastic wrap tightly over the open end of the can with the rubber band.
3. Tape one end of the straw onto the plastic wrap near the center.
4. Fold the cardboard so that it stands upright and extends at lease 3 cm above the top of the can.
5. Place the cardboard so that the free end of the straw just touches the front of the cardboard. Mark an X where the straw touches.
6. Draw three horizontal lines on the cardboard: level with the X, 2 cm above the X, and 2 cm below the X.
7. Position the cardboard so that the straw touches the X again. Tape the base of the cardboard in place so that it does not shift.
8. Observe the level of the straw at least once a day over a 5-day period. Record any changes.

Analysis and Conclusions

1. What factors affect how your model works? Explain your answer.

   Answers will vary. A typical answer would be that a rise in atmospheric pressure

   forces the plastic wrap down, which causes the straw to move upward on the

   cardboard. The reverse is true when the atmospheric pressure falls.

   _____

   _____

2. What does an upward movement of the straw indicate? a downward movement?

   a rise in atmospheric pressure; a fall in atmospheric pressure

   _____

   _____

   _____

   _____

3. Compare your results with the barometric pressures listed in your local newspaper. What kind of weather was associated with high pressure? with low pressure?

   Answers will vary. High atmospheric pressure generally indicates cooler, drier

   weather. Low atmospheric pressure generally indicates warmer, more humid

   weather.

   _____

   _____

M O D E R N   E A R T H   S C I E N C E

Chapter 24: Water in the Atmosphere

Small-Scale Investigation: Relative Humidity

Relative humidity is the ratio of the amount of water vapor in the air to the maximum amount of water vapor the air can hold at a given temperature. Using a relative humidity table, you can determine relative humidity.

Materials

thermometer; cheesecloth or light cotton material, about 5 × 5 cm; rubber band; water; small cardboard fan; relative humidity table on page 481 of your textbook

Procedure

1. Lay the thermometer on a table for a few minutes, until it adjusts to room temperature. Observe and record the dry-bulb temperature.
2. Secure the cheesecloth around the bulb of the thermometer with a rubber band. Moisten the cheesecloth with room-temperature water.
3. Hold the thermometer firmly at the top. Using your free hand, rapidly fan the bulb of the thermometer with the cardboard fan for 1 minute. Observe and record the wet-bulb temperature.
4. Repeat Steps 1–3 two more times. For each trial, subtract the temperature on the wet-bulb thermometer from the temperature on the dry-bulb thermometer. Use the table on page 481 of your textbook to determine the relative humidity for each trial. Calculate the average relative humidity for the three trials. Record your results.

Analysis and Conclusions

1. How do you account for the difference between the readings of the dry-bulb and wet-bulb thermometers?

   Some of the water from the cheesecloth evaporated and carried away heat from

   the thermometer, thereby reducing the temperature reading on the wet-bulb

   thermometer.

2. What is expressed by a relative humidity of 60 percent?

   A relative humidity of 60 percent indicates that the air is holding 60 percent of

   the amount of water it can hold at that temperature.

3. What might cause someone to feel hot and uncomfortable on a warm, humid day? Explain your answer.

   Answers will vary. A typical answer would be that the moist, humid air slows

   the rate of the evaporation of sweat from the skin. The evaporation of

   perspiration cools the body.

M O D E R N   E A R T H   S C I E N C E

Chapter 25: Weather

Small-Scale Investigation: Wind Chill

Wind chill is a term used by meteorologists to describe the relationship between temperature and wind speed. While values are usually read from a chart, wind chill is based on the physics of evaporative cooling. You can demonstrate the cooling power of moving air with a simple model.

Materials
shallow pan, 23 cm × 33 cm; water; thermometer; electric fan; clock or watch with second hand; metric ruler

Procedure
1. Place the pan on a level table, and fill it to a depth of about 1 cm with water at room temperature.
2. Lay the thermometer in the center of the pan with the bulb submerged. Be sure you can read the temperature without touching the thermometer.
3. Do not disturb the pan for 5 minutes, then read and record the water temperature.
4. Position an electric fan a few centimeters from the pan, and face the fan toward the pan. **Caution: Do not get the fan or cord wet.**
5. Turn on the fan at a low speed, and observe and record the water temperature every minute until the temperature remains constant.
6. If the fan has different speed settings, repeat Step 5 with the fan at a higher speed.

Analysis and Conclusions
1. How does the moving air affect the temperature of the water?

   The temperature of the water decreases.

   _____

   _____

   _____

2. The moving air is the same temperature as the still air in the room. What causes the temperature to change?

   Wind from the fan increases the rate of evaporation of the water; evaporation

   takes heat from the water, thereby decreasing the temperature.

   _____

   _____

   _____

3. On a cool, windy day, how would you dress to stay comfortable outdoors? Explain your answer.

   Answers will vary. A typical answer would be to minimize wind-chill effects by

   exposing little skin.

   _____

   _____

   _____

M O D E R N   E A R T H   S C I E N C E

Chapter 26: Climate

Small-Scale Investigation: Evaporation

Evaporation affects temperature. The converse is also true. You can demonstrate how temperature affects the rate of evaporation of water.

Materials

portable clamp lamp (or flexible-neck lamp) with an incandescent bulb; 3 small Petri dishes or watch glasses; 3 thermometers; 50-mL graduated cylinder; ring stand with 2 rings; clock or watch; meter stick; water

Procedure

1. Use the data table below to record your observations.
2. Assemble the ring stand on a table, with support rings at heights of 25 cm and 50 cm above the base. Position the lamp directly over the rings at a height of 75 cm.
3. Place a Petri dish or watch glass on the base of the stand, and place one on each of the two rings.
4. Lay a thermometer across each dish, and turn on the lamp. Wait 3 minutes, then read and record the temperature shown on each thermometer.
5. Remove the thermometers, and add 30 mL of water to each of the three dishes.
6. Make sure the dishes are lined up directly beneath the lamp, and keep the light on for 24 hours.
7. Turn off the lamp, and carefully pour the water from the first dish into the graduated cylinder and record any change in volume. Repeat this process for the other two dishes.

Analysis and Conclusions

1. At what distance from the lamp did the greatest amount of water evaporate? the least?

   The greatest amount of water evaporated at the highest temperature (closest to

   the light); the least amount of water evaporated at the coolest temperature

   (farthest from the light).

2. Explain the relationship between temperature and the rate of evaporation.

   The higher the temperature, the faster the water will evaporate.

3. Explain why puddles of water dry out much more quickly in summer than they do in fall or winter.

   The temperature is higher in the summer than it is in fall or winter, so the water

   evaporates at a faster rate.

Dish	Temperature	Amount of water evaporated
1		
2		**Answers will vary.**
3		

M O D E R N   E A R T H   S C I E N C E

Chapter 27: Stars and Galaxies
Small-Scale Investigation: Parallax

You can demonstrate the principle of parallax by viewing an object from several locations.

Materials
5 paper or plastic plates, about 15 cm in diameter (1 red, 4 blue); thread; scissors; masking tape; meter stick; ladder

Procedure
1. Cut five 1-m lengths of thread. Tape one end of each piece of thread to the edge of a paper plate.
2. Tape the free end of each piece of thread to the ceiling at various vertical heights about 30 cm apart in a random pattern with the red plate in front.
3. Stand directly in front of and directly facing the red plate at a distance of several meters.
4. Close one eye and sketch the position of the red plate relative to the blue plates in the background.
5. Take several steps back and to the right of your original position. Repeat Step 4.
6. Take several more steps directly back and make another sketch.
7. Repeat Step 6 once again.

Analysis and Conclusions
1. Compare your drawings. Did the red plate change position as you viewed it from different locations? Explain your answer.

   The plate does not actually change position, although it does appear to move

   when it is viewed from different locations.

2. What kind of results would you expect if you continued to repeat Step 6 at greater and greater distances? Explain your answer.

   The apparent change in position of the plate is greater at short distances than it

   is farther away because the angle formed by the observer's new line of sight and

   previous line of sight to the red plate decreases with distance.

3. If you noted the positions of several stars with a powerful telescope, what would you expect to observe about their positions if you sighted the same stars several months later? Explain.

   Answers will vary. A typical answer would be that the stars exhibit an apparent

   motion over a period of months that is similar to the apparent motion of the red

   plate in this investigation. In the case of the stars, the change in the observer's

   location is the result of the earth's orbit around the sun.

M O D E R N   E A R T H   S C I E N C E

Chapter 28: The Sun

Small-Scale Investigation: Solar Viewer

Solar telescopes enable astronomers to study the sun without looking directly at it. Using some common objects, you can make a solar viewer that functions somewhat like a solar telescope.

Materials

shoe box with lid; ruler; scissors; masking tape; 2 index cards; safety pin; piece of aluminum foil, about 4 cm × 4 cm

Procedure

1. Cut a hole with a diameter of about 3 cm in the center of one end of the shoe box. Tape an index card to the inside wall at the end of the box opposite the hole.
2. Tape the foil over the hole in the shoe box. Use the safety pin to make a tiny hole in the center of the foil. Put the lid on the shoe box.
3. Hold the box with the pinhole toward the sun. With your back to the sun, lift the lid and observe the image of the sun projected on the index card. Write down what you observe.
   Caution: Never look directly at the sun. Direct sunlight can damage your eyes.
4. Place a second index card inside the box at various distances from the pinhole. Write down what you observe at each distance.
5. Repeat Step 3 several times, making the pinhole slightly larger each time. Write down what you observe as the diameter grows larger.

Analysis and Conclusions

1. Does the image of the sun change in size or brightness as the distance between the pinhole and the index card is changed?

   Yes. As the distance between the pinhole and the card increases, the image

   becomes larger and fainter. As the distance decreases, the image becomes

   smaller and brighter.

   _____

   _____

2. What happens to the image as you make the pinhole larger?

   As the pinhole is made larger, the image becomes larger and less distinct. The

   edges of the image get more and more diffuse until the hole is so large that no

   clear image forms.

   _____

   _____

3. How is your solar viewer like a solar telescope? How is it different?

   The solar viewer works primarily because the sun is so bright. Like a solar

   telescope, the pinhole gathers light and projects an image. Unlike a solar

   telescope, the solar viewer does not magnify the projected image.

   _____

   _____

HRW material copyrighted under notice appearing earlier in this work.

M O D E R N E A R T H S C I E N C E

Chapter 29: The Solar System

Small-Scale Investigation: The Solar System

You can compare the relative sizes of the planets in the solar system and the distances among them with a simple model.

Materials

Planetary Data table on pages 600–601 of your textbook; drawing compass; metric ruler; notebook paper; scissors; sheet of paper, about 1.5 m²

Procedure

1. Use the table at right listing the sun and nine planets in the first column to record your measurements.
2. Study Table 29–1 on pages 600–601 of your textbook. Using the scale 1 cm equals 10,000 km, calculate the scale diameter and distance from the sun for each planet. Complete your table.
3. Using the information in your table and a drawing compass, draw a to-scale circle on a sheet of paper to represent each planet. Cut out each circle, and label it with the name of the planet it represents.
4. Out of the large sheet of paper, cut a circle that approximates the scale diameter of the sun.
5. Compare the relative sizes of the sun and the planets you have constructed.
6. Study the scale distance from the sun to each planet. Write a planetary-model plan that uses the scale distances you have calculated.

	Diameter (cm)	Distance from sun (cm)
Sun	130.0	—
Mercury	0.5	5,790
Venus	1.2	10,820
Earth	1.3	14,960
Mars	0.7	22,790
Jupiter	14.3	77,830
Saturn	12.1	142,700
Uranus	5.2	287,100
Neptune	5.0	449,700
Pluto	0.2	591,400

Analysis and Conclusions

1. Where is the greatest concentration of mass in the solar system?

   the sun

2. Would it be practical to complete your model using the scale 1 cm equals 100,000 km for both distance from the sun and diameter?

   No. The scale is too large to show the relative sizes of all the planets.

3. Compare the distances between the inner planets with the distances between the outer planets.

   The inner planets are closer together than the outer planets.

4. Why are models of the solar system that are displayed in classrooms often inaccurate?

   Most models do not accurately represent both the relative size of and the

   relative distance between the sun and all of the planets.

M O D E R N   E A R T H   S C I E N C E

Chapter 30: Moons and Rings

Small-Scale Investigation: Eclipses

As the earth and moon revolve around the sun, each casts a shadow into space. An eclipse occurs when one planetary body passes through the shadow of another body. You can demonstrate how an eclipse occurs by using clay models of planetary bodies.

Materials

modeling clay, sheet of notebook paper, penlight or small flashlight, metric ruler

Procedure

1. Make two clay balls, one about 4 cm in diameter and one about 1 cm in diameter.
2. Position the balls about 15 cm apart on the sheet of paper.
3. Turn off any nearby lights. Place the penlight approximately 15 cm in front of and almost on level with the large ball. Shine the light on the large ball. Sketch your model, noting the effect of the beam of light.
4. Reverse the positions of the two balls and repeat Step 3. Sketch your model, again noting the effect of the beam of light. You may need to prop up the smaller ball to center the shadow on the larger ball.

Analysis and Conclusions

1. What planetary body does the large clay ball represent? the small clay ball? the penlight?

   the earth; the moon; the sun

   _____

2. As viewed from the earth, what event did you represent with your model in Step 3? as viewed from the moon?

   As viewed from the earth, the model produced a lunar eclipse. As viewed from

   the moon, the model produced a solar eclipse.

   _____

3. As viewed from the earth, what event did you represent with your model in Step 4? as viewed from the moon?

   As viewed from the earth, the model produced a solar eclipse. As viewed from

   the moon, the model produced an eclipse of the earth.

   _____

4. What are the frequencies of the solar eclipse and the lunar eclipse in your model? How are the frequencies of these eclipses inaccurate? Explain your answer.

   In this model, there would be both a lunar and a solar eclipse each month. The

   eclipses would occur with this frequency because the model shows the earth

   and the moon orbiting in exactly the same plane around the sun. Actually, there

   is an angle of about 5° between the two orbits. As a result, one orbiting body is

   usually above or below the shadow of the other, and an eclipse does not occur.

   _____

Teacher's Guide to In-Depth Investigations

About In-Depth Investigations

Modern Earth Science In-Depth Investigations are designed to engage students in scientific thought and methods. The investigations reinforce concepts introduced in the textbook and provide students with an opportunity to employ scientific methods and to practice laboratory techniques. The *In-Depth Investigations* manual is a consumable book providing a worksheet format for each of the corresponding investigations found in the textbook. Each of the 30 investigations is a process-oriented laboratory intended to increase student enthusiasm and interest.

The *Teacher's Guide and Answer Keys to Investigations* contains a materials and equipment list, Teacher Notes, and annotated student worksheets. The Teacher Notes include a restatement of the objective and a list of the materials required for each investigation. The Prelab Discussion provides suggestions for introducing the investigation. The Teaching Strategies provide helpful suggestions to use during the investigation.

Contents for In-Depth Investigations

Notes on In-Depth Investigations

Chapter 1: Introduction to Earth Science

Investigation 1: Scientific Method

Approximate Time: 2 class periods

Objective: to use scientific method; to make predictions

Materials: meter sticks, hand lenses

Prelab Discussion: Before students begin the investigation, ask them to list a few scientific "facts" that have been proven incorrect. (For example, the sun travels around the earth, the earth is flat, nonliving things can produce living things, and mountains and valleys are formed by a shrinking core in the earth.) Ask students why detailed observations are important in their work. Have students list the procedures used in scientific method and discuss how scientific method can be used to solve problems in a logical way. Be sure students use the key terms of scientific method in their discussions.

Teaching Strategies

1. Students can work in pairs for this investigation. Before designating specific class periods for the investigation, explore areas around the school where observations can be made. If your school grounds are not suitable for the required observations, have students choose places near their home and record observations after school. Class time can be set aside for organizing data.

2. Caution students not to move or change anything near their investigation site. Emphasize to students that during this investigation they should use scientific method to develop their conclusions. Explain that reaching a conclusion without first gathering sufficient data can lead to incorrect results.

3. At the end of the investigation, each student or pair of students can explain to the class the steps they used to reach their conclusion. The class can then discuss if scientific method was used properly.

Chapter 2: The Earth in Space

Investigation 2: Earth–Sun Motion

Approximate Time: 2 class periods

Objective: to construct a shadow stick; to identify how changes in a shadow are related to the earth's rotation; to determine how the shadow stick can be used to measure time

Materials: magnetic compasses, metric rulers, pencils, clocks or watches, wooden boards, wooden dowels, masking tape

Prelab Discussion: Before students begin the investigation, be sure to describe the earth's period of rotation and relate it to the sun.

Discuss with students how people were able to "tell time" before the invention of mechanical sundials or electric clocks. Ask students to list some reasons why people would need to develop a method for telling time.

Teaching Strategies

1. In preparing the shadow sticks for this investigation, you may want to ask the industrial arts teacher to drill holes in the boards. Students should not be allowed to use a hand drill without adult supervision. If some students want to prepare the shadow stick at home using a drill, be sure to require that they written parental permission. An alternative to drilling holes in boards would be to have students use clay to hold the wooden dowels in place.

2. If wooden dowels are not available for making the shadow sticks, students may use unsharpened wooden pencils. Be sure that the shadow sticks are properly constructed before students begin the investigation.

3. You may wish to have students work in pairs for the investigation. Caution students not to look directly at the sun during the investigation.

4. You may wish to make a shadow stick and demonstrate to the class how to use it. Make measurements with the shadow stick during the times of day your classes meet. Make sure the students are aware of any adjustments to shadow stick height that may be required. Using the shadow stick to demonstrate, review the use of a directional compass.

5. You may wish to have a few students obtain data every 20 to 30 min. throughout an entire day. They can then report their findings to the class.

6. Extend the investigation by having interested students research the history of the sundial and describe to the class some of the more unique sundials that were used.

Chapter 3: Models of the Earth

Investigation 3: Contour Maps: Island Construction

Approximate Time: 2 class periods

Objectives: to interpret a contour map; to construct a three-dimensional clay model of an island

Materials: modeling clay, flat basins or large pans, scissors, water, plastic knives, wooden dowels, pencils

Prelab Discussion: Before students begin the investigation, you may want to review with them how to use a contour map. Explain that contour maps provide a view of the earth's surface from above. The contour lines on the map show elevation at the points connected by the contour lines.

Teaching Strategies
1. If available, dissecting pans can be used as replacements for the basins.
2. To save time, you may choose to make duplicates of the contour map on page 674 instead of having the students trace the map as instructed in Step 3 of the procedure.
3. You may wish to have students obtain a contour map of your area and then convert the contour map into a relief map.
4. Be aware that oils in the clay can stain papers or books.

Chapter 4: Plate Tectonics

Investigation 4: A Model of Convection Currents

Approximate Time: 1 class period

Objectives: to demonstrate convection current action; to show how convection currents may be the cause of plate motion

Materials: rectangular aluminum pans, cold water, beakers, ring stands, ring clamps, Bunsen burners with flame-spreader attachments, craft sticks, food coloring, medicine droppers, pencils, thermometers, lab aprons

Prelab Discussion: Point out to students that the driving force behind plate tectonics may be both large and hidden from view at the same time. To show students the difficulty of discovering the cause of plate motion, ask them whether drift could be the result of wind, waves, or other everyday phenomena. Remind them that volcanoes show that there is heat energy trapped underground. Suggest that the investigation may provide a possible explanation for how the earth's internal heat could drive the plates.

In the prelab preparation, students should diagram convection currents under a mid-ocean ridge. Have students refer to their textbooks or other sources to check their diagrams.

Teaching Strategies
1. This investigation is best performed by groups of two or three students.
2. Because of the volume of water used, you may want to set up the ring stands and tighten the beaker supports in advance to prevent water spillage.
3. If possible, pans should be set up no lower than 0.5 m below eye level.
4. Explain to students that the flame of the Bunsen burners must be controlled so that only a low, blue flame is used. Stress that the flames should be completely under the pan and should *not* climb up the sides of the pan. Check to make sure all students have properly adjusted flame heights before beginning the investigation.
5. Students may have difficulty in preventing the craft sticks from moving because of drafts, surface tension, or other classroom environmental disturbances. Try to eliminate as many sources of disturbance as possible.

Chapter 5: Deformation of the Crust

Investigation 5: Continental Collisions

Approximate Time: 1 class period

Objectives: to construct a model; to demonstrate with a model how the Himalayas formed as a result of the collision of India with Eurasia

Materials: thick cardboard, adding-machine paper, light-colored paper napkins, dark-colored paper napkins, wooden blocks, bobby pins, masking tape, metric rulers, scissors

Prelab Discussion: Have students describe a convergent boundary and a subduction zone. Stress the nature of the continental and oceanic lithosphere and their respective densities. Also, discuss the idea that the formation of new lithosphere at mid-ocean rifts requires the destruction of old lithosphere through the subduction process.

In the prelab preparation, students must draw a cross-sectional diagram of a plate collision. Have students refer to their textbooks or other sources to check their diagrams.

Teaching Strategies

1. This investigation is best performed by groups of two or three students.
2. You may want to cut the slits in the pieces of cardboard before students begin the lab. First use a sharp knife, then use a dull knife to widen the slits. If students cut the slits, demonstrate the procedure first with scissors, and emphasize safety precautions.
3. Emphasize that students must follow directions carefully as they set up their continental-collision models. Tell them to be sure that they secure the napkins with the bobby pins. If too much force is used, the bobby pins could suddenly spring forward.
4. Check to make sure that all of the students' models are set up correctly before they proceed with Step 7 in the procedure. Then, emphasize that in Step 7 students should pull on the paper strip slowly and carefully.
5. A few weeks prior to this investigation, you might want to obtain a map showing earthquake epicenters to be used in Extension 1. You can order a World Seismic Map from WARD'S (33 M 0028).

Chapter 6: Earthquakes

Investigation 6: Earthquake Waves

Approximate Time: 1 class period

Objective: to find the location of an earthquake's epicenter

Materials: drawing compasses, rulers, calculators

Prelab Discussion: Before students begin the investigation, use a world map to point out where the majority of earthquakes occur. Have students discuss the significance of the pattern in these occurrences. Ask why scientists must find the epicenters of earthquakes.

Teaching Strategies

1. Before students begin the investigation, be sure they understand the difference between the *focus* and the *epicenter* of an earthquake. The focus is the point on a fault at which movement occurs and an earthquake begins; the epicenter is directly above the focus and is where the surface or L waves begin to move.
2. Be sure students understand how to calculate the distances from each city to the epicenter of the earthquake in Step 2 of the procedure. These figures must be correct to accurately determine the epicenter of the earthquake on the map.
3. Emphasize to students that the circles on the map must intersect in order to determine the epicenter of the earthquake. If the circles do not intersect, tell students they must check their calculations in both prelab preparation and the procedure.
4. To extend the investigation, have interested students contact a nearby college or university to obtain seis-

mographic tracings. The tracings can be studied and discussed by the class.

Chapter 7: Volcanoes

Investigation 7: Hot Spots and Volcanoes

Approximate Time: 1 class period

Objectives: to construct a model; to use a model to demonstrate how the movement of the Pacific Plate is revealed by the orientation of the Hawaiian Islands and associated islands; to demonstrate the relationship between hot spots and volcanoes

Materials: cardboard, white poster board, plastic squeeze bottles, red-colored gelatin, scissors, eye droppers, metric rulers, sharp pencils, lab aprons

Prelab Discussion: Before students begin this investigation, ask them to name the major plates and their boundaries. Emphasize that the plates are moving relative to one another. Discuss convection currents and hot spots and their roles in adding volcanic material to the ocean floor.

Teaching Strategies

1. This investigation should be performed by groups of two or three students.
2. Prepare the gelatin a day before the investigation. Use cherry- or raspberry-flavored gelatin. One package should provide enough "lava" for at least two class periods.
3. Be sure the plastic squeeze bottles have screw tops with about a 3-mm opening at their tip.
4. Caution students that there must be very little air in their gelatin-filled dispensers. Too much trapped air could cause bursts of gelatin, leaving a mess on their papers. Relate the trapped air in the squeeze bottles and the bursts of gelatin to the lava activity of an erupting volcano.
5. Caution students that fruit gelatin may stain some types of fabric. You may want to have them wear aprons for this investigation. If any students do spill gelatin on their clothes, have them wash it out immediately with cold water and soap.
6. For the Extension section, you may wish to duplicate topographic maps of the floor of the Atlantic Ocean. Review the map with the students before they begin the Extension questions.

Chapter 8: Earth Chemistry

Investigation 8: Chemical Analysis

Approximate Time: 1 class period

Objectives: to use some of the techniques involved in chemical analysis; to determine whether water is an element or a compound

HRW material copyrighted under notice appearing earlier in this work.

39

Materials: lab aprons, 6-volt batteries, connecting wires, test tubes, electrodes, cork stoppers, beakers, water, Epsom salts, wood splints, matches, tongs or tweezers, graduated cylinders, safety goggles, masking tape, stirring rods, balances

Prelab Discussion: If students are not familiar with the process of electrolysis, you may to explain it before they begin the investigation.

In the decomposition of water by electrolysis, energy is transferred from the energy source to the decomposition products. A suitable electrolysis cell consists of two electrodes immersed in water. The electrodes are connected to a battery that supplies the electric energy to drive the decomposition reaction forward. One electrode, which is connected to the negative terminal of the battery, acquires a negative charge. This electrode is called the *cathode*. The other electrode, called the *anode,* is connected to the positive terminal of the battery. It acquires a positive charge. The electrolysis of water yields hydrogen and oxygen gases as a result of an oxidation-reduction reaction. Reduction occurs at the cathode, and hydrogen gas is the product. Oxidation occurs at the anode, and oxygen gas is the product.

Teaching Strategies
1. Epsom-salt solution is needed to speed up the electrolysis of water. It can be made by adding 1 part Epsom salt solution to 1 part water. Direct-current power units of 6 to 12 volts could be used if available. Six-volt dry cell batteries are recommended. Stainless steel electrodes, including leads, can be obtained from WARD'S (16 M 0501).
2. Students may have difficulty putting the inverted test tube into the beaker for the electrolysis of water. Demonstrate this technique beforehand, and have students practice the technique before beginning the investigation.
3. Be sure the students do not leave the electrolysis cell connected to the battery after the gases have been collected. In a closed room with a large number of cells operating past the time indicated in the instructions, the hydrogen content of the air could reach the explosion level. Properly ventilate the room during and after the experiment.
4. Make sure students understand the differences between elements, atoms, compounds, and molecules. Explain that water is a *compound* formed by chemically combining hydrogen and oxygen in a 2:1 ratio—2 hydrogen atoms to 1 oxygen atom. The smallest particle of water that still has the properties of water is a single *molecule* of water (H_2O). Hydrogen and oxygen are *elements* because they cannot be changed to a simpler form by chemical means. The smallest particle of hydrogen is a single hydrogen *atom*. The smallest particle of oxygen is a single oxygen atom.

Chapter 9: Minerals of the Earth's Crust
Investigation 9: Mineral Identification
Approximate Time: 2 class periods

Objective: to classify several mineral samples using a mineral identification key

Materials: copper pennies, glass squares, mineral samples, sand, steel files, streak plates, hand lenses

Prelab Discussion: Discuss with students ways in which minerals are classified. Have them cite the differences between a streak test and a scratch test. Discuss the variation of luster in minerals and the difference between cleavage and fracture. Ask why color is not the most reliable method for identifying minerals. (There is too much variation of color in minerals.)

Teaching Strategies
1. Before students begin the investigation, number the minerals they will be identifying. Be sure to use mineral samples that have properties related to the mineral identification key on page 45 of the student booklet. Record the names and numbers of the labeled minerals. You may wish to have students identify more than five samples. If Mohs hardness kits are available, you may wish to use them for this investigation.
2. Prior to class use, compare the mineral identification key with the mineral samples you have chosen to be certain that the key works effectively. A given identification key may not exactly describe any one mineral sample. You may find it helpful to look for any problems the students may have in identifying the minerals.
3. Review the mineral identification key with students before they begin the investigation.

Chapter 10: Rocks
Investigation 10: Classification of Rocks
Approximate Time: 2 class periods

Objectives: to identify various rock samples with the aid of a rock identification table

Materials: hand lenses, 10% dilute hydrochloride acid, medicine droppers, rock samples, safety goggles

Prelab Discussion: Before students begin the investigation, display a few different rock samples. Show at least one igneous, one sedimentary, and one metamorphic rock. Have students discuss how they would attempt to identify these rocks. Discuss with students some of the difficulties they may encounter in trying to identify rocks.

Teaching Strategies

1. Before students begin the investigation, number the rocks they will be identifying. Record the names and numbers of the labeled rocks.
2. Caution students to wear safety goggles during Step 4 of the Procedure and to use extreme care when working with the hydrochloric acid.
3. Review with students the rock identification table on page 51 of the student booklet. Point out significant characteristics of rocks that will help in identification. (For example, rock having fragments cemented together are usually sedimentary rocks.) Tell them to first identify the rock class to which each rock belongs and then try to name the rock.

Chapter 11: Resources and Energy

Investigation 11: Extraction of Copper from Its Ore

Approximate Time: 2 class periods

Objective: to extract copper from copper carbonate in a way similar to the way in which copper is extracted from malachite ore

Materials: test tubes, test-tube racks, test-tube clamps, copper (cupric) carbonate, iron filings, funnels, Bunsen burners, dilute sulfuric acid, water, safety goggles, lab aprons, heat-resistant gloves

Prelab Discussion: Before students begin this investigation, you may wish to discuss metallurgical processes. Metallurgy is the process whereby a metal is extracted from its ore and prepared for practical use. The desirable practical properties of many transition metals, such as copper, make them important to modern technology.

In their compounds, metals almost always exist in positive oxidation states. Therefore, to obtain a metal from its ore, the metal must be reduced (gain electrons). Ores that contain substantial amounts of impurities, such as rock, are often treated to concentrate the metal-bearing part and to convert some metal compounds into substances that can be more easily reduced.

Teaching Strategies

1. To make dilute sulfuric acid (3 molar), add concentrated sulfuric acid to water in the ratio of 1 part acid to 5 parts water. Caution students to be careful while working with sulfuric acid.
2. Copper carbonate, $CuCO_3$, is more readily available and more pure than malachite. Excellent results are obtained by using $CuCO_3$.
3. When copper oxide is redissolved in sulfuric acid (Step 6), the blue color indicates the presence of the hydrated copper (II) ion. There will probably be some CuO left in the test tube. If you wish to save it, you may direct students to wash it into a common container.

4. Make sure students understand what they are observing in Step 8. When the iron filings are added to the test tube containing copper sulfate, the copper ion is reduced to solid copper, which forms around the iron filings.

Chapter 12: Weathering and Erosion

Investigation 12: Soil Chemistry

Approximate Time: 1 class period

Objective: to identify different soil samples

Materials: soil samples (*A, B,* and *C* horizons), dilute hydrochloric acid, pH paper, water, medicine droppers, ammonia solution, test tubes, test-tube rack, cork stoppers, safety goggles, lab aprons

Prelab Discussion: Before students begin this investigation, emphasize that pedalfer soils are acidic; pedocal soils tend to be alkaline. Then show a pH color scale to the class. Explain that on this scale, each number lower than the preceding number represents 10 times the acidity. For example, a substance with a pH 5 is 10 times as acidic as a substance with a pH of 6.

Teaching Strategies

1. Be sure to review the safety guidelines for handling acids before starting this investigation.
2. If a wide-range pH test paper is not available, litmus paper could be substituted. If the litmus paper turns blue, pH is greater than 8.3; if it is red, it is less than 4.5; and if it is not affected, the pH would be somewhere between 4.5 and 8.3. Most soils will test in this middle range.
3. You may wish to furnish soil samples in order to have uniform results; or have students bring in samples from their own backyards.
4. Some students may have difficulty determining whether their soil samples are pedocal or pedalfer. Make sure students understand that each test is necessary to provide more information.
5. To emphasize scientific method, you may wish to have your students construct a table summarizing this investigation. The table should include the student's hypothesis, test results, and conclusion.
6. You may wish to have a class discussion about the relationship between rainfall and soil type. In the eastern half of the United States, rainfall exceeds 65 cm per year. The soil in this region is pedalfer. The soil is poor in soluble minerals, such as calcium carbonate, because the minerals are leached out by the rain. In the western half of the United States, excluding the Pacific Coast region, rainfall is less than 65 cm per year. The soil in this region is pedocal. The soil is rich in calcium carbonate because there is not enough rainfall to leach the mineral out of the soil.

HRW material copyrighted under notice appearing earlier in this work.

41

Chapter 13: Water and Erosion

Investigation 13: Sediments and Water

Approximate Time: 1 class period

Objective: to determine the erosional effect of water on different types of sediment

Materials: metric rulers, graduated cylinders, dry sand, dry silty soil, grease pencils, cardboard juice containers, large nails, timers, pans, water

Prelab Discussion: Before students begin this investigation, explain that soil is composed of rock material, water, and organic matter. The rock material in soil can be clay, sand, or silt. Emphasize that these materials are classified by particle size.

Have students hypothesize which type of soil—sandy or silty—would erode more rapidly. Have them explain their reasoning. (There is more runoff from silty soil because it cannot absorb as much water as sand and because water does not flow as rapidly through it.)

Teaching Strategies
1. This investigation is best performed by groups of two or three students.
2. If natural sediments are not readily available, play sand and potting soil can be used.
3. If there are sediment slopes near the school, you may wish to take students on a field trip. Have them observe the effect of rainfall on the local sediment. If you are in an area of farms, reinforce the idea that contour plowing and/or terracing are techniques used to control slope stability where oversaturation would cause soil erosion. A visit with a local farmer might provide a personal perspective on soil-erosion phenomena.

Chapter 14: Groundwater and Erosion

Investigation 14: Porosity

Approximate Time: 1 class period

Objective: to measure and compare the porosity of three samples that represent rock particles

Materials: beakers, graduated cylinders, 4-mm plastic beads, 8-mm plastic beads

Prelab Discussion: Before students begin the investigation, be sure they understand how porosity affects the amount of water a rock can hold. Discuss how sorting influences the porosity of rocks. Emphasize that the size of particles is not important in sorting. Point out that sediments composed of all large particles can have the same porosity as sediments composed of all small particles. A poorly sorted sediment contains particles of all sizes.

Teaching Strategies
1. During this investigation students will record porosity as a decimal and as a percentage. You may wish to have students practice calculating percentage problems.
2. As students are filling the beaker with the plastic beads, instruct them to shake the beaker gently. This will ensure that the beads are as closely packed as possible.
3. Be sure students understand that the volume of the pore space is equal to the volume of water than can be added to the beads.
4. Before students begin the Extensions section, you may wish to review the terms *permeability* and *capillarity*. Explain that permeability refers to the ease with which water flows through the open spaces of a rock or sediment. Capillarity refers to the attraction of water molecules to other materials, such as soil. Have students describe the difference between porosity and permeability. Ask students to name materials that have a high permeability and a low porosity. (For example, limestone, which contains many interconnected cracks.) Ask what materials have a high porosity and a low permeability. (For example, clay, which has irregular particles that do not have connected pore spaces.)
5. Extend this investigation by having different groups of students work with soil, sand, sorted pebbles, unsorted pebbles, or combinations of the materials. Have the groups then calculate the pore spaces of these substances.

Chapter 15: Glaciers and Erosion

Investigation 15: Glaciers and Sea Level

Approximate Time: 2 class periods

Objective: to construct a model; to use the model to simulate what would happen if the Antarctic polar ice sheet melted

Materials: shallow pans, sand, small pebbles, water, milk cartons, metric rulers, wooden blocks, freezer

Prelab Discussion: Have students describe the formation and growth of a glacier. Discuss how temperature and the balance between snowfall and melting ice affect the growth of a glacier. Describe continental glaciers and point out that they are found only in Greenland and Antarctica. Emphasize that large quantities of fresh water are held in glaciers.

For the prelab preparation, you may wish to draw other figures on the chalkboard to give students extra practice in calculating area.

Teaching Strategies

1. This investigation is best performed by groups of two or three students.
2. Be sure students use pans large enough to keep the melting ice from overflowing. Although various pan sizes may be used, students should use the proportions of ice, water, sand, and pebbles indicated in the investigation.
3. Because the ice may not completely melt during class time, it may be necessary to store the pans in an elevated position until the next class period. You may want to make provisions for this ahead of time.
4. Sand will clog sink drains. Tell students to drain the pan carefully and to properly dispose of the sand when the investigation is finished.
5. Before students begin the Extension section, you may wish to review metric-system conversions and the proper method for dividing unit exponents. You may also wish to have students practice percentage problems.
6. To extend the investigation, discuss methods that scientists are developing to utilize some of the fresh water that is locked in polar ice sheets. For example, one idea is to have icebergs towed from Antarctica to areas around the world where fresh water is needed. Have students discuss the benefits of developing these methods and the obstacles that may be encountered.

Chapter 16: Erosion by Wind and Waves

Investigation 16: Beaches

Approximate Time: 2 class periods

Objective: to examine a model showing how the forces generated by wave action build up, shape, and erode beaches

Materials: masking tape, small pebbles, sand, stream tables or large plastic containers, small rocks, water, large wooden blocks, milk cartons, plaster of Paris, metric rulers

Prelab Discussion: Before students begin the investigation, ask them to explain why shorelines are the most rapidly changing areas on the earth. Have them describe some of the results of wave erosion on a shoreline. Discuss how longshore currents are formed and how they affect a shoreline.

Teaching Strategies

1. This investigation is best performed by groups of two or three students.
2. After students have made their plaster blocks in the prelab activity, tell them the plaster must be completely dry before they peel away the milk cartons.

3. You may wish to have your first-period class set up the stream tables, and then use the same water throughout the day with your other classes. Have the last class of the day clean up the materials. If stream tables are not available, large shallow pans can be used.
4. Tell students to pour the water slowly into the stream tables to prevent alteration of the shoreline.
5. Emphasize to students that the waves they generate in the stream tables should be a consistent size.
6. Extend the investigation by having students set up barrier islands in the stream tables. Have students build the barrier islands at several distances from their model shorelines. Students can then generate waves and describe what they observe.

Chapter 17: The Rock Record

Investigation 17: Fossils

Approximate Time: 1 class period

Objective: to use various methods to make models of trace fossils

Materials: soft carbon paper, hard objects (shell, key, paper clip, etc.), leaves, model clay, newspaper, plaster of Paris, plastic containers, plastic spoons, tweezers, water, pencils or wooden dowels, white paper, lab aprons

Prelab Discussion: Before students begin the investigation, have them describe how the remains of organisms become fossilized. Have students categorize the means of fossilization into those that preserve the organism and those that preserve traces of the organism.

Teaching Strategies

1. Stress to students that the plaster mixture must be the correct consistency (heavy cream). If the plaster is too watery, it will take a long time to harden.
2. Remind students that all plaster must be removed from the containers before the end of the class period. If the plaster remains in the containers overnight, the plaster will be nearly impossible to remove. Caution students not to dispose of plaster in the sink.
3. While the plaster is hardening, you may want to display examples of assorted fossils. Have students make drawings of the actual organisms that have been fossilized and try to identify each organism. Tell students to write a description of each organism's environment.

Chapter 18: A View of the Earth's Past

Investigation 18: Geologic Time

Approximate Time: 1 class period

Objective: to convert geologic time into a frame of reference that is easier to understand

HRW material copyrighted under notice appearing earlier in this work.

43

Materials: meter sticks, adding-machine paper, fine-point pens or sharpened pencils, reference books with information on the geologic eras

Prelab Discussion: Have students discuss why tracing geologic time as far back as the estimated beginnings of the earth, 4.6 billion years ago, is such a difficult task. Tell students that, like a novel, rocks tell us a story, but unlike the novel, many parts are missing. Discuss how scientists search the rock and fossil records for events in geologic time in order to understand our present world better.

Teaching Strategies

1. For this investigation, have students work on the floor so that they can lay down the full length of the adding-machine paper.

2. Some students may have difficulty converting geologic time into meters, centimeters, and millimeters. You may wish to help students by doing a few conversion problems on the chalkboard. Use 1 m to represent 1 billion years.

   1 billion years   = 1 m
   600 million years = 0.6 m, 60 cm, or 600 mm
   10 million years   = 0.01 m, 1 cm, or 10 mm
   1 million years   = 0.001 m, 0.1 cm, or 1 mm

3. Remind students that they should always measure their distances from the TODAY line. Emphasize that their lines should be made with a fine-point pen or sharp pencil.

4. Tell students that in Step 5 of the procedure, they may find varying dates given in different reference books for the events listed. Accept all reasonable dates that students use to indicate these events on their scales. You may wish to add other geologic events to those listed in Step 5. Events occurring over long periods of time are especially effective in helping students grasp the meaning of geologic time.

5. When students have finished their paper scales, have them add to the scales by listing common animals or plants that existed during the different periods or epochs. For example, students could list *Pterosaurs* under the Jurassic Period.

6. Extend the investigation by having each student make a time scale of his or her life. They should include any important events that have occurred. You may wish to have them extend the time scale into the future, based on their predictions and imaginations.

Chapter 19: The History of the Continents

Investigation 19: History in the Rocks

Approximate Time: 1 class period

Objective: to discover how the geologic history of an area can be determined from the arrangement of fossils and rock layers

Materials: paper, pencils

Prelab Discussion: Before students begin the investigation, briefly review important concepts and terms from Chapter 17, including relative age, the law of superposition, nonconformities, and index fossils. Emphasize the importance of index fossils to paleontologists trying to determine relative ages of rock layers.

Teaching Strategies

1. Tell students to study Figures 19.1 and 19.2 carefully before they begin the investigation. Be sure students notice that the fossils shown in *Mollusca* and *Chordata* in Figure 19.1 are grouped together because these fossils span the length of the geologic period indicated. They can be placed in any order within these periods.

2. Stress that a time span of one million years is a relatively short period of time in the earth's total geologic history.

3. You may wish to display a chart of the geologic eras and have students discuss the differences and similarities in life-forms found during the different eras. Have them point out similarities to present-day life-forms.

4. Have interested students research other index fossils and the geologic periods in which the organisms existed. Students can then develop their own fossil arrangements, such as those shown in Figure 19.2. Students may also have to research to expand the fossil reference chart in Figure 19.1. Ask students to present their results to the class and have classmates interpret them.

Chapter 20: The Ocean Basins

Investigation 20: Ocean-Floor Sediments

Approximate Time: 1 class period

Objective: to determine the relationship between the size of sediment particles and their settling rate in water

Materials: graduated cylinders, dry mixed sand/soil, grease pencils, sieves (4 mm, 2 mm, 0.5 mm), stopwatch, water, paper cups, clear plastic columns, rubber stoppers, adhesive tape, ring stands and clamps, balances, teaspoons, paper towels, metric rulers

Prelab Discussion: Before students begin the investigation, discuss how ocean-basin sediments differ. Have students describe the different types and sizes of sediments that are found in the deep ocean, on the continental shelves, and near the shore. Ask students to describe some sources of ocean sediment. Discuss how particle size in sediment affects the sediment's settling rate in water.

Teaching Strategies

1. Students should work in pairs for this investigation.
2. Provide a container with a trash-can liner for disposal of wet soil samples.
3. In the prelab procedure, remind students to place a container underneath the 0.5 mm sieve in which to collect the fine particles. You may wish to have the sieving completed ahead of time.
4. Stress that time measurements should be as accurate as possible throughout the investigation. Tell students that in Step 2, they should time the first measurement as soon as soil is poured into the clear plastic column. Plastic columns can be obtained through WARD'S (36 M 4193).
5. You may wish to have students research types of mineral sediments that are found on the ocean floor and, if applicable, describe the economic importance of these minerals.

Chapter 21: Ocean Water

Investigation 21: Ocean Water Density

Approximate Time: 1 class period

Objective: to observe the effects of temperature and salinity on the density of salt water

Materials: beakers, Bunsen burners, modeling clay, graduated cylinders, grease pencils (yellow and red), heat-resistant gloves, plastic straws, ring stands, ring clamps, safety goggles, table salt, teaspoon, wire gauze, thermometers, metric rulers, scissors, freezer, water

Prelab Discussion: Discuss with students the effect of salinity in ocean water as fresh water enters the ocean from large rivers and heavy rainfall. Explain to students that ocean water is approximately 96.5 percent pure water and 3.5 percent dissolved salts. Ask them if slight variations in salinity change the density of ocean water.

Teaching Strategies

1. Students may have difficulty balancing their hydrometer. Stress that they should try to keep the hydrometer vertical in the water to obtain an accurate reading.
2. Caution students to wear safety goggles. You may wish to review the safety guidelines for heating before students light their Bunsen burners. Caution them to wear heat-resistant gloves while holding the thermometers in the water.

3. Explain to students that salt lowers the freezing point of water. Ask students why the ability of salt to lower the freezing point of water is invaluable to people living in regions where heavy snowfall is common.
4. Have several students research the effects of pollution on the density of ocean water. Have these students report their findings to the class.

Chapter 22: Movements of the Ocean

Investigation 22: Wave Motion

Approximate Time: 1 class period

Objectives to simulate wave motion; to observe how energy generates wave motion in water; to observe the properties of waves

Materials: colored pens or pencils, cloth ties (about 50 cm in length), graph paper, markers, meter sticks, sheets of paper (2 m × 1 m), thin rope (2.5 m long)

Prelab Discussion: Before students begin this investigation, ask them to describe the causes of the ocean's movements. Explain that winds generate waves and that gravitational forces of the moon and the sun cause tidal motion. Have students describe the circular motion of water particles within a wave. Review properties of waves by having students describe the crest and trough of a wave, wave height, wave length, and wave period.

Teaching Strategies

1. On the chalkboard, review with students how to calculate wave speed.
2. If you cannot obtain rolls of paper that are at least 1 m wide, you can substitute large pieces of newsprint taped together.
3. Stress to students that the waves they generate with the rope should be of a similar size and have a constant motion.
4. Have each student team practice Step 5 a few times before marking the crests and troughs on the paper.
5. You may wish to demonstrate that objects generally move very little on the water's surface. Set up a pan of water and place a cork on the surface. Then generate small waves using a wooden block. Point out to students that the cork does not move in the direction of the wave but just follows the up-and-down movement of the water.

HRW material copyrighted under notice appearing earlier in this work.

45

Chapter 23: The Atmosphere

Investigation 23: Air Density and Temperature

Approximate Time: 1 class period

Objective: to observe how the density of air is affected by changes in its temperature

Materials: beakers, Celsius thermometers, cold water, hot water, glycerin, ice, disposable syringes (60 cc), petroleum jelly

Prelab Discussion: Before students begin this investigation, review the relationships between temperature and the volume of a gas, and between volume and the density of a gas. Emphasize to students that the temperature corresponding to a reported gas density must be specified because density changes when a gas is heated or cooled.

Teaching Strategies

1. You may wish to have students calculate the following practice density problems:
   a. A block of magnesium has a mass of 14.3 g and a volume of 8.5 cm^3. What is the density of magnesium: (1.7 g/cm^3)
   b. The density of iron is 7.87 g/cm^3. What is the mass of 10 cm^3 of iron? (78.7 g)
2. Have students wear safety goggles during this investigation. Tell them to hold their fingers over the syringe cap when pushing down the syringe plunger. This will keep the cap from accidentally shooting across the room and possibly hurting someone.
3. Before students begin Step 4, remind them that they are not to stir the ice with their thermometers. This could result in the breaking of the bulb. Also remind them not to let the thermometer touch the bottom or sides of the beaker or the water temperature measurement will be inaccurate.
4. Because some students may have difficulty plotting points on the graph in Step 8, you may want to review graphing skills briefly.
5. Make sure students understand that the mass of the air in the syringe remains constant throughout the investigation.
6. To extend this investigation you may wish to discuss *Charles' law* with students. The relationship that exists between the volume of a gas and its temperature is summarized by this law.

Chapter 24: Water in the Atmosphere

Investigation 24: Relative Humidity

Approximate Time: 1 class period

Objective: to use wet-bulb and dry-bulb thermometer readings to determine relative humidity

Materials: cotton cloth, plastic containers, ring stands with rings, string, Celsius thermometers, water

Prelab Discussion: Before beginning the investigation, ask your students to define *relative humidity* and discuss the different ways relative humidity can be measured. Then ask your students to describe how a psychrometer works. Review Graph 24.1 (on page 110 of the student booklet) with students to help them determine how much moisture air can hold at a given temperature.

Teaching Strategies

1. Some students may have difficulty setting up the apparatus. Have them study Figure 24.1. You then may wish to demonstrate the correct way to set up the apparatus.
2. To ensure that students understand the relationship between the dry-bulb temperature reading and the wet-bulb reading, have them complete the following statement: "As the difference in temperature between the dry and wet bulbs increases, relative humidity [increases, *decreases*]." Some students may have difficulty understanding this concept. You may wish to review pages 480 and 481 in the textbook with students.
3. Tell students that Table 24.1 has been worked out by actual experimentation.
4. You may wish to obtain permission to have a group of students perform this investigation outside the school building. They can then compare the relative humidities of the classroom with the relative humidity outside. Alternatively, they may compare classroom humidity to that in the cafeteria or locker rooms.
5. As an extension of this investigation, give students a list of various types of environments, such as a desert, an island off the coast of southern California, or a farm in the midwestern United States. Have students compare the relative amounts of water vapor in the air over each area. During a class discussion, have students give reasons for their choices.

Chapter 25: Weather

Investigation 25: Weather Map Interpretation

Approximate Time: 1 class period

Objective: to study the symbols used on a weather map; to gain an understanding of the relationships between temperature, pressure, and winds

Materials: paper, pencil, colored pencils

Prelab Discussion: Before students begin this investigation, review the map legend on page 116 of the student booklet with the class. Have students explain what each symbol represents. Explain to them that the symbols and numerals on weather maps are used to make predictions about the weather.

Teaching Strategies

1. Some students may have difficulty interpreting the weather map at first. You may wish to ask each student or lab group if they need any instructions clarified, and then go through a few steps in the procedure with the class.
2. Remind students that weather systems move from west to east in North America. Using a weather map from a daily paper, have students check the weather in western cities, and then forecast the next day's weather for various cities in the east. Have students record their forecast for each city on a chart similar to Table 25-1 on page 117.
3. To extend the investigation, have students watch a weather report on the television news. Then have them compare weather symbols used on maps on television with weather symbols used on maps in newspapers.
4. As an additional extension of this investigation, have students look at a current weather map in a local newspaper. Ask students to forecast the next day's weather in their town or city. Discuss why students made those particular predictions. Have them check the accuracy of their forecasts in the next day's paper. Most likely, some students used different methods when developing their forecasts. Discuss which methods worked best.

Chapter 26: Climate

Investigation 26: Factors That Affect Climate

Approximate Time: 1 class period

Objective: to explore how the angle of the sun's rays and the distribution of land and water affect climate

Materials: tape, black construction paper, containers, flashlights, graph paper, heat lamps, meter sticks, Celsius thermometers, soil, water

Prelab Discussion: Before students begin the investigation, ask them to describe how these factors affect climate. Remind students that the latitude of a region determines how much solar energy the area receives, and that land and water absorb and release heat at different rates. You may want students to list the other factors that affect climate. (Answers should include topography, ocean currents and wind patterns, and seasonal winds.)

Teaching Strategies

1. Caution students to be careful when using the heat lamp. The lamp may get very hot.
2. In Step 4, tell students to count only the completely lit squares on the graph paper. To speed the counting of the lighted squares in Steps 4 and 5, have students

draw the largest rectangle possible within the lighted area that they traced. Have them calculate the number of squares by multiplying the width of the rectangle by the length rather than by counting each square individually. Remind students to count only the whole squares remaining in the traced area.
3. Remind students that thermometers are fragile and should be handled with care. Tell students that they should notify you immediately if a thermometer breaks. Caution students not to touch the mercury if it spills on the table or floor.
4. Before allowing students to go on to Step 8 of the procedure, instruct them to use a pencil to loosen a small area of the soil to a depth of a little more than 0.5 cm. Have them place a thermometer bulb in the loosened soil, then gently pack the soil around it. Warn the students that they must never dig into unloosened soil with the thermometer.
5. You may wish to have students graph the results they obtain in Step 9 of the procedure. Students could plot time (minutes) along the x-axis and temperature (°C) along the y-axis for both water and soil on the same graph.

Chapter 27: Stars and Galaxies

Investigation 27: Star Magnitudes

Approximate Time: 1 class period

Objectives: to determine the effect of distance on brightness and the relationship between brightness and color

Materials: 3-volt flashlight bulbs, AA batteries, plastic-coated wires with stripped ends (15 cm and 20 cm), masking tape, paraffin bricks (12 × 6 cm), aluminum foil (12 × 12 cm), large rubber bands, rulers, desk lamps with incandescent bulbs

Prelab Discussion: Before students begin this investigation, you may want to review squaring numbers. Ask students to explain the difference between absolute magnitude and apparent magnitude. Students should be able to explain why stars that appear to have equal brightness are not always the same distance from the earth. Also discuss with students the reason the flame of a candle ranges from blue, at the center of the flame, to white-yellow, and then orange at the edges. (The flame is hottest at the center, around the burning wick.)

Teaching Strategies

1. Because of the amount of equipment used in this investigation, have students work in groups of two or three.
2. To prepare the wire to be used by the students, use a pair of electrician's wire strippers. If they are unavailable, use a single-edged razor blade or mat knife to cut the wires to the proper lengths, then use the blade to strip the plastic covering from the ends

of the wire, exposing about 1.5 cm of bare wire. Although it may save time to have each group of students strip their own wire, having them use a blade is not recommended for obvious safety reasons.

3. To increase the stability of the flashlight, you may want to have students tape the completed flashlight to a sturdy strip of cardboard.

4. Avoid doing the investigation at midday. At that time, the sun may be too bright to sufficiently darken the room. You may, however, place the apparatus inside a large cardboard box to shield it from room light.

Chapter 28: The Sun

Investigation 28: Size and Energy of the Sun

Approximate Time: 1 class period

Objective: to perform a simple experiment from which the sun's diameter can be calculated; to collect energy from sunlight and estimate the amount of energy produced by the sun

Materials: shoeboxes with lids, aluminum foil, safety pins, index cards, metric rulers, thin sheet metal (2 × 8 cm), flat-finish black paint, desk lamps with 100-W bulb, glass jars with lids, clay, Celsius thermometers, masking tape, scissors, pencils

Prelab Discussion: Before beginning this investigation, caution students never to stare directly at the sun. You may want to review with your students the concept of ratios. Point out that energy from the sun reaches the earth in the form of electromagnetic radiation or light waves. Have students explain how the way in which the sun produces heat energy is different from the way in which burning wood produces heat energy. (In the sun, energy is released when nuclei fuse to form heavier nuclei. There is no substance in the sun being burned.)

Teaching Strategies

1. It is suggested that you prepare the jar lids yourself or have them prepared by the industrial arts department. To prepare the lids, use a power drill to make a hole in the center of the lid. The hole should have a diameter slightly larger than that of the thermometer. If you choose to allow the students to make the holes, have them wear safety goggles. They should place the jar lids on a solid wooden plank placed on the floor to protect the floor from damage. Then they should punch the hole using an awl and hammer.

2. You may want to prepare strips of sheet metal prior to class by carefully cutting them from empty aluminum soda cans.

3. Instruct the students on the proper way to handle a thermometer. Remind them never to force the thermometer through a hole that is too small. Also tell them not to pinch the sheet metal too hard around the base of the thermometer or the thermometer bulb will be crushed.

4. This investigation is most easily conducted by groups of two or three students.

5. In Step of the prelab preparation, check to be sure that all the thermometers are attached securely to the jar lids.

5. As an extension of this investigation, have students research the different types of solar collectors and the way in which they function.

Chapter 29: The Solar System

Investigation 29: Crater Analysis

Approximate Time: 1 class period

Objectives: to experiment with making craters; to discover the effect of speed and projectile angle on the formation of craters

Materials: meter sticks, plaster of Paris, marbles, large marbles, water, shoe boxes, protractors, scissors, toothpicks, masking tape, markers, safety goggles, lab aprons

Prelab Discussion: Before students begin this investigation, have them name objects in the solar system (other than the earth's moon) that have craters. Ask them if they know of any craters in the United States. Explain that Mercury, the earth's moon, and many other moons in the solar system are covered with craters because they have little or no atmosphere and, therefore, have no direct source of mechanical and chemical weathering. Weathering of the earth's surface and the process of plate tectonics have caused the disappearance of many of the craters that once covered the planet.

Teaching Strategies

1. Review the importance of wearing safety goggles to protect the eyes from any splashing plaster or from a stray marble. Remind students to drop, not throw, the marbles into the plaster.

2. To save time you may want to prepare all the plaster of Paris yourself and then distribute it to the students. If students prepare the plaster themselves, be sure they work on spread newspaper. Caution students not to pour any of the plaster of Paris into the sink.

3. Students should conduct this investigation in groups of two or three.

4. You may want to have students practice using a protractor. Draw various angles on a sheet of paper. Hand out copies and have the students use a protractor to measure and record the angles.

5. You may want to display some photos of the moon's surface. Discuss the differences in the features of the craters and have students hypothesize about the possible causes of these differences. You also may want to display photos of Mercury and ask them to identify the features of the planet. As an additional extension, have students compare the surface features of Mercury and the moon.

Chapter 30: Moons and Rings

Investigation 30: Galilean Moons of Jupiter

Approximate Time: 2 class periods

Objective: to verify that the orbital motions of Jupiter's moons obey Kepler's third law

Materials: metric rulers, calculators

Prelab Discussion: Before students begin this investigation, it is very important that they understand Kepler's third law. Have students identify the variables and the constant in the equation. Review the use of exponents in recording very large numbers. Point out that Jupiter's four largest moons are called the Galilean Moons because they were discovered by Galileo Galilei and were used by him in support of a heliocentric solar system.

Teaching Strategies

1. Students can do this investigation without a partner. However, if you do choose to have the students work in groups of two, try to pair a student proficient in math with one who is not as adept.

2. Have students practice reading the chart in Figure 30.2. Give them different dates and an approximate time of day, and have them identify the positions of the four moons at that time.

3. If students have trouble working with exponents, give them a list of large numbers that they can practice converting into exponential form.

4. Check the students' results in Step 2 of the procedure before allowing them to continue. Point out that if they do not calculate p^2 and r^3 carefully, they will arrive at the wrong value of K.

5. After students have completed Step 8 of the procedure, check their values for p^2 and r^3 before allowing them to continue.

6. Although students will get various values for K, these values will all be approximately 300. Explain that this difference in values results from using numbers that have been rounded off.

7. To extend the investigation, have students with access to a telescope observe Jupiter and its moons on five consecutive nights. Tell them to make illustrations of what they were able to observe. Most astronomy books have celestial charts from which you can find Jupiter's position on a specific night. Instruct students where to look in the sky.

M O D E R N   E A R T H   S C I E N C E

Chapter 1: Introduction to Earth Science
In-Depth Investigation: Scientific Method

Objective
In this investigation, you will use scientific method to predict a change in a small area of the environment.

Prelab Preparation
Explain the difference between a quantitative observation and a qualitative observation.

In a quantitative observation, measurements and other statistical information are
recorded such as mass, time, or temperature. Accuracy is important in this type of
observation. The information learned in a quantitative observation is then applied
to a qualitative observation, which is a summary of the factual information that has
been learned. In a qualitative observation, an opinion is expressed such as water
is denser than air or shadows in the Northern Hemisphere are longer in the winter.

Observations
1. Describe the observation you chose to interpret?

   Answers will vary. _____

   _____

2. Describe your results. Was your prediction correct? Explain. If it was not correct, describe
   what factors may have affected the results, and how you can improve your observations or
   revise your predictions.

   Answers will vary. _____

   _____

   _____

Analysis and Conclusions
1. Which of your senses did you use most to make your observations? How could you improve
   observations using this sense?

   Answers may vary, but sight should be very important. _____

   _____

2. What could you have used to measure, or put into numbers, many of your observations? Is
   quantitative observation better than qualitative observation? Explain.

   Answers will vary. Students should be able to see the value in using both

   quantitative and qualitative observations. _____

   _____

3. Can inferences generally be relied on as true? Explain.

No. Inferences are only the beginning of scientific method. They are

based on observations and are used to interpret how the observations

came to be. They are not based on fact.

4. If your predictions are found to be incorrect, was the act of forming your inferences a waste of time? Explain.

No. Scientists may pose many inferences before they find one that can be

verified. During the process scientists will learn something new each time.

This will aid them in their search for supportable conclusions.

5. After reporting the results of a prediction, how might a scientist continue his or her research?

The scientist will need to make many more observations until a conclusion

about the validity of any prediction can be certain. The original inference

may then be accepted, changed somewhat, or rejected entirely.

Extensions

1. Decide whether each statement is an inference (I) or an observation (O).
   a. Grass is present inside the puddle. __O__
   b. The grass surrounding the puddle is greener and taller than that inside the puddle. __O__
   c. During a rainstorm, some soil is washed into the puddle. __O__
   d. Water always runs downhill. __O__
   e. Gravity causes water to run downhill. __I__
   f. The soil that washes out of the puddle will eventually become part of a stream. __I__
   g. Brownish water contains suspended soil particles. __O__
   h. The soil particles are suspended because water is flowing fast. __I__
   i. When the rain stops, the puddle water looks clear. __O__
   j. Mud cracks result from drying the soil. __I__
2. Briefly describe your experiment and how you tested it.
   Answers will vary.

M O D E R N   E A R T H   S C I E N C E

Chapter 2: The Earth in Space

In-Depth Investigation: Earth-Sun Motion

Objective

In this investigation, you will construct a shadow stick in order to identify how changes in a shadow are related to the earth's rotation. You will also determine how the shadow stick can be used to measure time.

Observations

1. Why is it necessary to draw a line on the notebook paper showing the direction north?

   So that all shadow sticks are oriented in the same direction.

2. Is the line connecting the dots on the paper a straight line?

   No; it is an arc.

3. In what direction did the shadow move?

   east

4. **a.** What is the length, in centimeters, of the first shadow?

   Answers will vary.

   b. What is the length, in centimeters, of the last shadow?

   Answers will vary.

5. What distance, in centimeters, did the shadow move in 30 minutes?

   Answers will vary.

Analysis and Conclusions

1. In what direction did the sun appear to move in the 30-minute period?

   east to west

2. In what direction does the earth rotate?

   west to east

3. If you made your shadow stick half as long, would its shadow move the same distance in 30 minutes? Explain.

   No. The stick would cast a shorter shadow and this shadow would not move

   as much as a longer shadow.

4. How might a shadow stick be used for telling time?

If properly oriented, the shadow stick can be used to tell time based on the same principle as a sundial. There must be equally divided units marked on the base of the shadow stick. The units can be established after a total arc has been made by the shadow stick, and can then be used as a standard of time.

Extension

Repeat this investigation at different hours of the day, such as early in the morning, early in the afternoon, and early in the evening. Record the results and any differences that you observe. How can shadows be used to tell whether it is morning or afternoon? How can shadow sticks be used to tell direction?

Answers will vary. Students should observe that a west or northwest shadow indicates morning and an east or northeast shadow indicates afternoon. To use a shadow stick for the direction students should observe that in the morning, shadows point west; at noon, shadows point north; in the afternoon, shadows point east.

M O D E R N   E A R T H   S C I E N C E

Chapter 3: Models of the Earth
In-Depth Investigation: Contour Maps—
Island Construction

Objective
In this investigation you will use a contour map to construct a three-dimensional clay model of an island.

Analysis and Conclusions
1. What is the contour interval of your map?

   10 m

2. How could you tell the steepest slope from the gentlest slope from observing the spacing of the contour lines?

   The contour lines on the steepest slope are close together. This indicates a

   rapid increase in elevation over a short distance. The contour lines on

   the gentlest slope are spaced further apart. This indicates a gradual increase

   in elevation over a longer distance.

3. What is the elevation above sea level for the highest point of your model?

   Above 60 m but lower than 70 m

4. How do you know if there are any points on your model that are below sea level? If there is such an area, where is this location and what is its elevation?

   Yes, near point A; the elevation of the depression is at least −10 m but not

   greater than 20 m below sea level.

5. What feature is located at *C* on your model?

   A pass between two hills

6. What is the elevation of point *B* on your model?

   Between 20 m and 30 m above sea level.

7. Is there a bench mark? If so, what is the elevation?

   Yes. Above 60 ft. but lower than 70 ft.

Extension
Based on observations of your model, what conclusions can you make about where people would live on this island? Explain.

Most people would likely live in the upland regions of the south central part of the

island where the higher elevations would be less affected by flooding due to high

tides or storm surges. The gentler slopes would also be easier to cultivate and

build on.

Laboratory Notes

M O D E R N   E A R T H   S C I E N C E

Chapter 4: Plate Tectonics
In-Depth Investigation: A Model of Convection Currents

Objectives
In this investigation you will demonstrate convection-current action. You will attempt to show how convection currents may be the cause of plate motion.

Prelab Preparation
Make a cross-sectional diagram to show what convection currents under a mid-ocean ridge might look like.

Observations
1. What was the motion of the food coloring in the water before the Bunsen burners were lighted?

    It spreads out randomly.

2. Describe any differences in the movement of the food coloring just after the burners were lighted compared with earlier movements.

    Food coloring rose toward the surface more than it moved laterally.

3. Describe the motion of the food coloring when the sticks began to move.

    Food coloring rose quickly toward the surface and spread toward the ends

    of the pan.

4. What is the relationship between the motion of the sticks and the motion of the water?

    The same

5. Record the thermometer readings.

Center thermometer _____ Answers will vary _____

End thermometer #1 _____ Answers will vary _____

End thermometer #2 _____ Answers will vary _____

6. Describe how the temperature of the water relates to the motion of the food coloring in the water.

The food coloring showed that water was rising because of heating by the

Bunsen burners. As the water traveled to the sides of the pan it mixed

with cooler water and the water's temperature was lowered, and it sank.

_____

Analysis and Conclusions

1. In relation to plate tectonics, what do the Bunsen burners, water, and craft sticks represent?

Bunsen burners are the source of heat for mantle convection (asthenosphere).

Water is the equivalent of plastic mantle rock. Craft sticks are the equivalent

of the lithospheric plates.

2. Explain how the motion of the water affected the motion of the craft sticks.

The water movement pushed the sticks in the direction is was moving

through convection.

3. Does the temperature of the water at different places relate to the flow pattern that you observed with the food coloring? Explain.

Yes; there is the high heat flow from mid-ocean ridges.

_____

_____

Extensions

1. Suggest some reasonable modifications of the investigation that would make it a more realistic illustration of the process of seafloor spreading.

Use a deeper pan to sustain the convection flow; the liquid could have been

more viscous (syrup or oil); use more sticks. The sticks could be color

coded to show magnetic field reversals; new sticks could be added as the

old ones move away.

2. Find a diagram that shows measurements of heat flow from ocean crust. Do the readings show a pattern of high heat flow at the mid-ocean ridges?

Yes; there is also high heat flow from volcanic hot spots that lie away from

mid-ocean ridges.

_____

M O D E R N   E A R T H   S C I E N C E

Chapter 5: Deformation of the Crust
In-Depth Investigation: Continental Collisions

Objectives

In this investigation, you will create a model to help explain how the Himalaya Mountains formed as a result of the collision of India into Eurasia.

Prelab Preparation

Make a cross-sectional diagram to show what happens when a plate carrying oceanic and continental crust collides with a plate carrying continental crust at its edge.

Observations

1. What happens as the dark-colored paper napkin first comes in contact with the stationary block of wood?

   It begins to form folds.

2. What happens to the dark-colored and light-colored paper napkins as you continue to pull on the paper strip?

   The dark-colored paper napkin continues to form folds and the light-colored

   napkin begins to form folds on top of them.

Analysis and Conclusions

1. Explain what is represented by the dark-colored napkins, the light-colored napkins, and the wood blocks.

   oceanic crust; continental margin; continental crust

2. What plate tectonics process is represented by the motion of the paper strip in the model? Explain.

<u>This model shows an oceanic plate being subducted under a continental</u>

<u>plate. It also shows a mountain-building process.</u>

3. What type of mountains would result from the kind of collision shown by the model?

<u>Folded mountains</u>

4. Explain differences between the model and the real Himalaya Mountains.

<u>Faulting is present in the Himalayas and earthquakes cannot be simulated in</u>

<u>the model.</u>

Extensions

1. Study a map of earthquake epicenters. Describe the pattern of epicenters in the Himalayan region. Does the pattern suggest that the Himalayas are still growing?

<u>Shallow earthquakes occur in the Himalayas and intermediate quakes occur</u>

<u>on the north side of the range. The pattern suggests continued growth</u>

<u>of the Himalayas.</u>

2. Read about the breakup of Gondwanaland and the movement of India toward the northern hemisphere. Write about stages in India's movement. List the time at which each important event occurred.

<u>About 200 million years ago, India begins to move northward; about 65</u>

<u>million years ago, India passes over a hot spot and the Deccan Plateau</u>

<u>lavas are formed; about 50 million years ago, India collides with Eurasia.</u>

M O D E R N   E A R T H   S C I E N C E

Chapter 6: Earthquakes
In-Depth Investigation: Earthquake Waves

Objective
In this investigation, you will find the location of an earthquake's epicenter by applying a method similar to one that scientists use. You will also use a model to demonstrate the difference between two kinds of earthquake waves.

Prelab Preparation
1. How long would it take P waves moving at 6.1 km/s to travel 100 km? How long would it take P waves to travel 200 km?

   16.4 seconds; 32.8 seconds

2. How long would it take S waves moving at 4.1 km/s to travel 100 km? How long would it take S waves to travel 200 km?

   24.4 seconds; 48.8 seconds

3. What is the time lag between the arrival of P waves and S waves over a distance of 100 km? What is the time lag for a distance of 200 km?

   8.0 seconds; 16.0 seconds

Part I Observations
1. Record your observations in Table 6.1.

Table 6.1 Epicenter Distances

City	Lag time (seconds)	Distance from city to epicenter
Austin	150 s	1,875 km
Bismarck	168 s	2,100 km
Portland	120 s	1,500 km

2. Find the epicenter of the earthquake using the map in Figure 6.2, page 34.

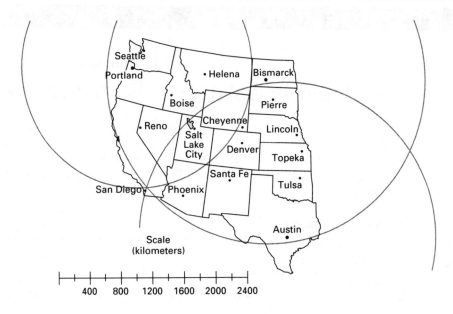

Figure 6.2

Analysis and Conclusions

1. The location of the earthquake epicenter is closest to what city?

   San Diego

2. Why must there be measurements from three different locations to find the epicenter of an earthquake?

   To ensure that the location is accurate. Sometimes the first two circles intersect in more than one place. The third circle intersects both of the other two circles in only one place.

Extensions

1. What is the probability of an earthquake occurring in the area where you live?

   Near California's San Andreas Fault, the Pacific Northwest, or Alaska the probability is high. Elsewhere in the United States, the chances are much lower.

2. If an earthquake did occur in your area, what would be its probable cause?

   Other than in fault or rift zones, an earthquake could result from old buried faults or the rebounding of land, such as in New England after the ice retreated during the last ice age.

M O D E R N   E A R T H   S C I E N C E

Chapter 7: Volcanoes

In-Depth Investigation: Hot Spots and Volcanoes

Objective

In this investigation you will construct a model to demonstrate how the movement of the Pacific Plate is revealed by the orientation of the Hawaiian Islands and associated islands. You will also demonstrate the relationship between hot spots and volcanoes.

Prelab Preparation

How many individual volcanoes are shown on the map?

At least 40 individual volcanic mountains

Observations

1. **a.** What is the relationship between the amount of gelatin deposited on paper and the height of the volcanoes?

    The more gelatin deposited, the higher the volcano

   b. Does this relationship seem to apply for the real Emperor seamounts and Hawaiian Islands?

    Yes. The Hawaiian Islands appear to have a greater volume.

2. **a.** Was the flow of gelatin smooth or was it accompanied by little bursts? What caused the bursts of gelatin? How is this similar to the behavior of magma?

    Trapped air caused the gelatin bursts. This is similar to trapped gases in

    magma, which cause explosive volcanic eruptions.

   b. How does the behavior of the gelatin compare to the behavior of lava that is building the island of Hawaii?

    Volcanoes on Hawaii are a quiet, flow type with only minor fountaining.

    The extrusion of gelatin is like a quiet eruption.

Analysis and Conclusions

1. Using your model for reference, label Midway Island, Kauai, and Hawaii on the appropriate lines in Figure 7.2. Then label the Emperor Seamounts and Hawaiian Islands.

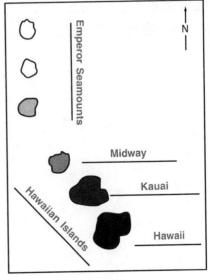

Figure 7.2

1. Describe the direction of motion of the Pacific plate over the last 70 million years as demonstrated by the model.

   At first the movement was NNW. Then, 40 million years ago, the movement

   became northwesterly.

2. The hot-spot model represents only a small number of volcanoes that exist in the Emperor Seamount chain and the Hawaiian Islands. Based upon the count of volcanoes you completed in the prelab preparation, does it seem that the hot spot now under the Hawaiian Islands has been active for at least 70 million years? Explain.

   Yes, there do not seem to be any gaps in the line of volcanoes.

3. The distance from the southernmost tip of the island of Hawaii to Midway Island is about 2,400 km.

   a. What is the rate of motion of the Pacific plate (in cm/yr), assuming that the plate traveled that distance in 40 million years?

   240,000,000 cm in 40,000,000 yrs = 6 cm/yr.

   b. Geologists estimate that the current rate of motion of the Pacific plate is 2 cm/yr. If this rate is different from the rate that you determined, what is the explanation for the difference?

   The Pacific plate seems to have slowed in its northward motion.

   c. Explain the possible relationship between the rate of plate motion and the size of the volcanic islands in the Hawaiian Islands.

   It may be that the slowing of plate motion allows more time for lava to

   build larger islands

Extension

Obtain a map of Atlantic Ocean seafloor topography to answer the following. Both Iceland and the Azores Islands are located over the Mid-Atlantic Ridge. What are the differences and similarities between these islands and the Hawaiian Island chain?

Iceland is on top of a portion of the Mid-Atlantic Ridge and does not seem to be

part of a line of islands. The Azores are not part of a line of islands. They are all

volcanic islands made of lava.

M O D E R N   E A R T H   S C I E N C E

Chapter 8: Earth Chemistry
In-Depth Investigation: Chemical Analysis

Objective
In this investigation, you will use some of the same techniques chemists use to determine the composition of matter. You will test to determine whether water is an element or a compound.

Prelab Preparation
Do you think water is a compound or an element? How could you find out?

Compound; try to separate it into simpler parts

Observations
1. a. What evidence do you see that a reaction is taking place?

    Bubbles slowly form around each electrode.

   b. Is the speed of the reaction the same at each electrode?

    No; one is producing more bubbles than the other.

   c. Which electrode, positive or negative, gave off more gas?

    Negative

2. a. What happens when you bring a flame near the mouth of the test tube that was over the negative electrode?

    It makes a popping noise.

   b. What gas is in the test tube?

    Hydrogen

3. a. What happens when you thrust a glowing splinter into the test tube that was over the positive electrode?

    The splint bursts into flame.

   b. What gas is in the test tube?

    Oxygen

Analysis and Conclusions

1. Is water a compound or an element? Explain.

   A compound; water can be broken down by a chemical reaction into two

   simpler components.

2. Are hydrogen and oxygen compounds or elements? Explain.

   Elements; elements cannot be broken down into simpler substances.

Extensions

1. a. Baking soda and some antacid tablets fizz when put into water. Is this a physical or a chemical change? How do you know?

   Chemical change; a new substance, a gas, is formed.

   b. How might you be able to prove it?

   Perform tests to identify the new substance.

2. a. When substances such as sugar or salt dissolve in water, is the resulting substance a mixture or a compound? How do you know?

   Mixture; no new substance is formed.

   b. How might you be able to prove it?

   Allow the water to evaporate. The original sugar or salt will remain.

M O D E R N   E A R T H   S C I E N C E

Chapter 9: Minerals of the Earth's Crust
In-Depth Investigation: Mineral Identification

Objective
In this investigation, you will classify several mineral samples using a mineral identification key.

Prelab Preparation
1. Are all sand grains the same color? Is it possible to only use the property of color to identify sand grains?

   No, more information is needed because the grains are different colors.

2. What is the hardness of a mineral sample that is scratched by a copper penny but not by a fingernail?

   3

Observations
Record your observations in Table 9.2.

Table 9.2 Mineral Tests

Sample Number	Color/ Luster	Hardness	Streak	Cleavage/ Fracture	Mineral Name
1					
2	Answers will vary depending on mineral samples.				
3					
4					
5					

Analysis and Conclusions
1. Identify your mineral samples. Describe the properties that helped you identify each one.

   Answers will vary depending on mineral samples used.

   _____

   _____

   _____

   _____

2. Although color is the most obvious property of a mineral, it is difficult to identify a mineral by its color alone. Explain.

The color of a mineral may vary from sample to sample due to impurities.

3. a. What is the difference between a scratch test and a streak test?

A scratch test is a test for hardness; a streak test is a test for color.

A streak will rub off but a scratch will not.

b. Why do some very hard minerals leave no streak?

Streak depends on the mineral being in a finely powdered form.

Some minerals are too hard to leave a powdered streak on porcelain.

4. Which minerals have the same color as their streak? Which do not?

Answers will vary.

Extensions

1. Diamonds and graphite are both made of the element carbon, but they are not considered the same mineral. Explain.

Diamonds and graphite have very different properties, especially hardness.

2. Corundum, rubies, and sapphires all have the chemical formula Al_2O_3, and they are considered the same types of mineral. Explain.

Corundum, rubies, and sapphires all have the same hardness, crystal

shape, etc. They vary only in color and crystal size.

M O D E R N   E A R T H   S C I E N C E

Chapter 10: Rocks
In-Depth Investigation: Classification of Rocks

Objective
In this investigation, you will will use a rock identification table to identify various rock samples.

Observations
Record your observations in Table 10.2.

Table 10.2 Rock Descriptions

Specimen	Description of Properties	Rock Class	Rock Name
	Answers will vary.		

Analysis and Conclusions
1. What properties were most useful in identifying each rock sample?

   Texture, structure, and mineral composition. _____

   _____

   _____

2. Were there any samples that you found difficult to identify? Explain.

Answers will vary. Students will most likely find fine-grained rocks,

such as shale, most difficult to identify.

3. Were there any characteristics common to all the rock samples?

Answers will vary. All the rocks contained minerals and were hard.

4. How can you distinguish between a sedimentary rock and a foliated metamorphic rock when both have observable layering?

Many sedimentary rocks are made of layers of compressed and cemented

sediment particles. Metamorphic rocks usually have bands of minerals.

Extensions

1. Name properties that distinguish the following pairs of rocks from one another.
   a. granite and limestone:

   Granite has mostly visible mineral crystals; limestone contains calcite, which

   reacts with hydrochloric acid, and individual crystals that are not visible.

   b. obsidian and sandstone:

   Obsidian is made of volcanic glass with a smooth surface; sandstone has

   a rough, sandy surface.

   c. pumice and slate:

   Pumice has many holes and looks spongy; slate splits into thin plates.

   d. conglomerate and gneiss:

   Conglomerate has rounded fragments; gneiss has a banded appearance.

2. How many rocks did you collect that were igneous? How many were sedimentary rocks? How many were metamorphic rocks? Try to name the types of rocks you collected.

Answers will vary.

M O D E R N   E A R T H   S C I E N C E

Chapter 11: Resources and Energy

In-Depth Investigation: Extraction of Copper from Its Ore

Objective

In this investigation, you will extract copper from copper carbonate in much the same way that copper is extracted from malachite ore. You will also demonstrate a method of copper purification.

Prelab Preparation

1. Define the term *electrolysis* as it applies to chemistry.

   The conduction of electricity through an ionic solution,

   together with the resulting chemical changes

2. Define the term *weathering*.

   Process in which rocks are broken up and decomposed

Part I Observations

1. What color is copper carbonate?

   Green to blue

2. After you heat the copper carbonate, what color is the substance in the test tube?

   Black

3. a. What is the new compound formed in the bottom of the test tube?

   Copper oxide

   b. Does the new compound take up as much room in the test tube as did the original copper carbonate? Explain.

   No; because CO_2 gas is given off during the reaction, there is

   less solid material left in the test tube.

4. What is the color of the liquid solution formed when dilute sulfuric acid is added to the test tube?

   Blue

5. a. What indicates that a chemical reaction is taking place?

   Bubbles rise from the solution. A red solid forms.

   b. If there is a change in the iron filings describe the change.

   They become a red solid.

   c. If there is a change in the color of the solution, describe the change.

   The solution becomes less blue.

6. In your demonstration, what do the iron filings represent in the actual process of extraction of copper from its ore?

   Cast iron scrap metal

Analysis and Conclusions

1. a. Disregarding any condensed water on the test-tube walls, what is the new substance formed in the first test tube called?

Copper oxide

_____

b. Does the new substance take up as much space in the test tube as did the copper carbonate? Explain.

No; because CO_2 gas is given off during the reaction, there is less

solid material left in the test tube.

_____

2. a. When the iron filings were added to the second test tube, what indicated that a chemical reaction was taking place?

Bubbles rose from the solution.

_____

b. Describe any change to the iron filings.

The iron filings became red in color.

_____

c. Describe any change in the solution.

The solution became less blue.

_____

3. In the actual process of extracting copper from its ore, the copper sulfate solution is allowed to flow over cast iron scrap metal. The loose layer of copper that forms on the scrap metal is then separated and pressed into bars or redissolved for purification. What do the iron filings represent in the actual process of extracting copper from its ore?

The filings represent iron scrap metal.

_____

Extension

Suppose that a certain deposit of copper ore contains a minimum of 1.0% copper by mass and that copper sells for 30 cents per kilogram. Approximately how much could you spend to mine and process the copper out of 100 kg of copper ore and remain profitable?

Less than 30 cents

MODERN EARTH SCIENCE

Chapter 12: Weathering and Erosion
In-Depth Investigation: Soil Chemistry

Objective
In this investigation, you will identify a soil sample as being pedocal or pedalfer.

Prelab Preparation
1. Define the following terms.

   a. acidic Having a pH of less than 7; capable of reacting with
   a base to form a salt

   b. alkaline Having a pH of more than 7; capable of reacting with
   an acid to form a salt

   c. neutral A substance that is neither acidic or alkaline

2. Refer to page 61 to find the pH for each of the following substances. Then, classify the following substances as acid, alkaline, or neutral.

   a. soft drink _____acid_____   **f.** lemon juice _____acid_____

   b. pure water _____neutral_____   **g.** ammonia _____alkaline_____

   c. milk of magnesia _____alkaline_____   **h.** milk _____neutral_____

   d. orange juice _____acid_____   **i.** vinegar _____acid_____

   e. sea water _____alkaline_____   **j.** blood _____alkaline_____

3. What is the pH of the tap water?
   generally 6–8

Observations
1. **a.** What is the pH of the topsoil sample?
   Depends on soil sample, usually 5 to 9

   b. Is this acid or alkaline?
   Depends on soil sample; below 7—acid; above 7—alkaline; 7—neutral

2. **a.** What is the pH of the subsoil sample?
   Depends on soil sample

   b. Is this acid or alkaline?
   Depends on soil sample

3. Based on your pH results, hypothesize about whether your soil sample is pedalfer or pedocal.
   If acid, probably pedalfer; if alkaline, probably pedocal.

4. **a.** How many of the rock particles were silicates?
   Depends on soil sample

   b. How many were calcium carbonate?
   Depends on soil sample

5. What color is the liquid above the soil sample? Draw the contents of your test tube in test tube *a* shown below. Label each layer of material.

   Depends on soil sample

6. Is iron present in your soil sample? Draw the contents of your test tube in test tube *b* shown below.

   Depends on soil sample; If liquid is red, there is iron present.

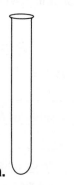 

a.   b.

Analysis and Conclusions

1. Is your soil sample pedalfer or pedocal? Explain your answer based on the results of the tests performed in Steps 4–6 of the procedure.

   If acid, and if silicates and iron are present, pedalfer.

   If alkaline, and carbonates are present, pedocal.

2. **a.** What type of soil, pedalfer or pedocal, would you treat with acidic substances to help plant growth? Explain why.

   Pedocal. Pedocal soils tend to become too alkaline to support plants.

   b. Explain why the acidic substances in the item above are usually spread on the surface of the soil.

   Plants grow mostly in the topsoil or A horizon. Also, rainwater tends

   to leach the substances down through the layers of soil.

Extension

Why has the use of phosphate and nitrate detergents been banned in some areas?

The addition of waste water containing phosphate and nitrate

detergents caused so much algae growth that waterways were

overgrown.

M O D E R N   E A R T H   S C I E N C E

Chapter 13: Water and Erosion

In-Depth Investigation: Sediments and Water

Objective

In this investigation you will determine the erosional effect of water on different types of sediment.

Prelab Preparation

1. Describe and explain your observations of the water you pinched between your thumb and index finger.

   The water droplets cling to the fingers and to each other. This happens

   because of the water molecules' attraction for each other and the

   fingers. However, when the fingers are moved apart past a certain distance,

   the water no longer clings to the water on each finger.

2. Does your wet finger or dry finger slide more easily along the table's surface? Explain why.

   On a non-porous surface, the wet finger; water acts as a lubricant.

Observations

1. What was the volume of water poured into container A? What was the volume poured into container B?

   Container A _____ Answers will vary. _____

   Container B _____ Answers will vary. _____

2. Which sediment held more water?

   The sand

3. How long did it take for the water level to drop to the surface of the sand? How long did it take the water level to drop to the surface of the silty soil?

   Sand _____ Answers will vary. _____

   Silty soil _____ Answers will vary. _____

4. Through which sediment was the water able to flow faster?

   Sand

5. Record your measurements of mound heights in Table 13.1.

Table 13.1

Sediment	Mound height (cm)
Dry sand	
Dry soil	
Wet sand	
Wet soil	

6. Which mound of dry sediment is higher, the soil or the sand?

Sand

7. Which mound of wet sediment is higher?

Sand

8. Is either wet mound higher than it was when dry? Explain why.

Both were higher. Water's ability to cohere to other water

molecules and to other objects in contact with it holds the

sediment together.

9. What happens as more water is added to the sediments? Explain why.

The mounds collapse. This happens because water also acts as a

lubricant so that after a certain amount of water has been added,

the sediment particles slide apart and are separated by the water.

Analysis and Conclusions

1. Based on your answers to Observations 2 and 4, would water erode an area of silty soil or an area of sand more quickly? Explain.

Silty soil. Because water can soak into and flow more quickly

through the sand, there is less runoff to cause erosion. The silty soil,

however, neither holds as much water as sand nor allows the rapid

flow of water. Thus, more of the silty soil will be carried away

by runoff.

2. Would a hillside covered with sand or a hillside covered with silty soil be more easily eroded during moderate rainfall? Explain with reference to Observations 6 and 7.

The hillside of silty soil. When slightly damp, the cohesive

effect of water holds the sand together better than

it holds the soil. The silt is more likely to slide downhill.

3. Would a hillside covered with sand resist erosion during an extended period of heavy rain? Would a hillside covered with silt resist the erosion? Relate your answer to Observation 9.

Although the sand would withstand the rains better than the

silt, eventually the lubricating action of the water would

cause large masses of the sediment to flow downhill and wash away with

the runoff. Silty soil may erode in the form of a mudslide.

Extension

Describe ways in which slopes covered with soil can be made more resistant to erosion.

Plant grass or other plants, build tiers of barricades, reduce

the steepness of the slope. Other answers are acceptable.

M O D E R N   E A R T H   S C I E N C E

Chapter 14: Groundwater and Erosion
In-Depth Investigation: Porosity

Objective
In this investigation, you will measure and compare the porosity of three samples that represent rock particles.

Prelab Preparation
What is a *meniscus*?

The curved surface of a liquid in a graduated cylinder. Volume readings

should be taken at the meniscus.

_____

Observations

1. What is the volume of the beaker?
   Answers will vary depending on the size of the beaker.

2. Do the 7-mm beads represent well-sorted large rock particles, well-sorted small rock particles, or unsorted rock particles?
   Well-sorted large rock particles.

3. What is the total volume of the large beads, including the pore space?
   Answers will be the same as the volume of the beaker.

4. How much water did you add to the beaker?
   Answers will vary.

_____

5. What is the volume of the pore space between the large beads?
   Answer will be equal to the volume of water added.

6. What is the porosity of the large plastic beads? (Give your answer as a decimal and also as a percent.)
   Answers will vary. Porosity = answer (Step 5)/answer (Step 3)

7. Do the 4-mm plastic beads represent well-sorted large rock particles, well-sorted small rock particles, or unsorted rock particles?
   Well-sorted small rock particles.

8. a. What is the volume of the pore space between the small plastic beads?
      Answer will be equal to the volume of water added.

   b. What is the porosity of the small beads?
      Porosity = answer (Step 9)/answer (Step 3)

9. Do the equal volumes of the small and large beads represent well-sorted or unsorted rock particles?
   Unsorted rock particles

_____

Analysis and Conclusions

1. Compare the large-bead porosity with the small-bead porosity.

   Their porosities are equal.

2. Does porosity depend on particle size? Explain.

   No; even though large particles have larger spaces between them,

   the particles themselves take up a larger amount of space. Thus,

   the increased size of the pores is balanced by the increased

   space taken up by the particles.

3. What effect did mixing the bead sizes have on the porosity? Explain this effect.

   The porosity decreased because the pore spaces between the large

   beads is filled by the small beads.

Extensions

1. Each of the three graphs in Figure 14.1 represents one of the following properties plotted against particle size: porosity, permeability, and capillarity. Particle size is plotted on the vertical axis, increasing from bottom to top. Porosity, permeability, and capillarity are plotted on the horizontal axis, increasing from left to right.

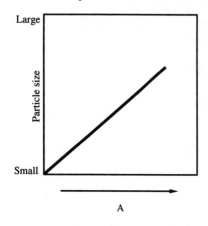
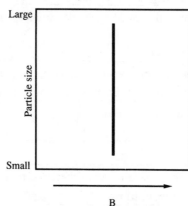
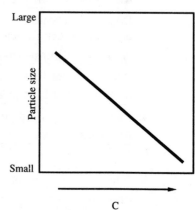

A   B   C

Figure 14.1

    a. Which graph represents porosity? Explain.

    Graph B. Porosity does not change with particle size.

    b. Which graph represents permeability? Explain.

    Graph A. The larger particle size, the greater the permeability.

    c. Which graph represents capillarity? Explain.

    Graph C. The smaller the particle size, the greater the capillary.

2. What would be the effect on the porosity of coarse gravel if it were mixed with fine sand? Describe the experiment you conducted to find the answer.

   The porosity would decrease. The pore space of the gravel

   would be partly filled by the fine sand, thus reducing

   the porosity.

M O D E R N   E A R T H   S C I E N C E

Chapter 15: Glaciers and Erosion

In-Depth Investigation: Glaciers and Sea Level

Objective

In this investigation, you will construct a model to simulate what would happen if the Antarctic ice sheet melted.

Prelab Preparation

What is the area of Figure 15.1a? What is the volume of Figure 15.1b?

$a = 2$ cm $\times$ 4 cm = 8 cm^2; $b = 3$ cm $\times$ 3 cm $\times$ 2 cm = 18 cm^3

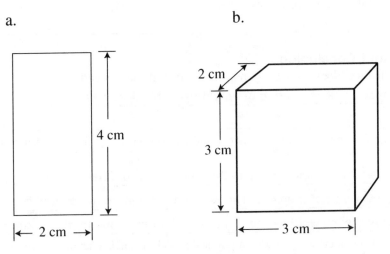

a.   b.

Figure 15.1

Observations

1. What is the surface area of the bottom of the pan?

   Answers will vary.

2. What is the volume of the ice block? What is the area of one side?

   Answers will vary depending on the size of the container.

3. What is the depth of water at the deepest point in the pan?

   Answers will vary.

4. What is the distance from the end of the pan covered with sand to the point where the sand touches the water?

   Answers will vary.

5. What percentage of the total area of the pan is covered by ice?

   Answers will vary.

6. **a.** As the ice begins to melt, what happens to the bottom of the ice block?

   Sand and pebbles are becoming embedded in the ice.

   b. What happens to the sand under the ice block?

   The sand is being smoothed out by the ice block.

7. What is the calculated rise in water level in the pan?

Answers will vary depending on measurements.

8. a. After the ice is completely melted, what is the depth of water in the pan?

Answers will vary but the water level will be deeper.

  b. What is the difference in the water level after the ice has melted?

Answers will vary but the water level will be higher.

  c. How does this compare with the value you calculated in Step 7? Explain any differences.

Answers will vary. Differences between calculated and measured values can

occur because of poor technique or variables unaccounted for.

9. a. After the ice has melted, what is the distance from the end of the pan covered with sand to the point where the sand touches the water?

Answers will vary.

  b. What is the difference in this distance after the ice has melted?

Answers will vary.

Analysis and Conclusions

1. How is the ice block model different than a real glacier on earth?

A real glacier moves due to gravity; the ice block remains stationary.

2. How does this model represent what would happen on the earth if the Antarctic ice sheet melted?

If the Antarctic ice sheet melted, the sea level would rise and the ocean

would cover part of the land. The Antarctic continent uncovered by the

melting of the ice would show evidence of erosion due to glaciation.

3. During this investigation, you used a physical model to simulate an occurence in nature. How else do scientists use models? What kinds of errors may occur when using models?

Scientists use models to represent something that cannot be observed.

However, the models are based upon properties that have been observed.

Any model can have errors because no model can be an exact duplicate

of nature.

Extension

What is the area of the earth, in square kilometers, that is covered with water?

358,000,000 km²

What is the average worldwide rise in sea level, in kilometers, if the Antarctic ice sheet melted? Convert this answer to meters.

approximately 0.050 km; approximately 50 m

How is the mathematical model for the rise in sea level similar to the physical model used earlier in this investigation?

Both models can be used to predict what will happen in nature.

M O D E R N E A R T H S C I E N C E

Chapter 16: Erosion by Wind and Waves
In-Depth Investigation: Beaches

Objective
In this investigation, you will examine a model showing how the forces generated by wave action build up, shape, and wear away beaches.

Part I Observations
1. What happened to the beach when water was first poured into the container?

   Fine sand and some pebbles were washed down into the water.

2. a. Describe the appearance of the beach after the wave action has taken place.

   The beach has less sand, more pebbles, and appears rougher.

   b. What happened to the particles of fine sand?

   They moved into deeper water.

3. What will happen to the beach if it has no source of additional sand?

   It will turn into a pebble beach.

4. What happened to the beach when you generated waves in front of a breakwater?

   Only a small part of the waves got through, and they disappeared

   quickly. The movement of sand was reduced to practically nothing.

Part II Observations
5. a. What effect would a longshore current have on sand just offshore?

   The sand would move with the current.

   b. How would this affect the sand on the beach?

   The sand on the beach would move into the water to take the

   place of the sand moved by the current.

6. What happened when you generated waves after placing a jetty in the sand?

   Sand built up behind the jetty. Some sand remained in front

   of the jetty.

Analysis and Conclusions
1. a. How does wave action build up a beach?

   Waves form beaches by wearing away rock fragments and depositing

   them along a shoreline.

b. How does wave action wear away a beach?

Waves wear away beaches by carrying away sand.

_____

2. How do longshore currents change the shape of a beach?

Longshore currents move sand along the shore.

_____

3. What effect would a series of jetties have on a beach?

A series of jetties would tend to preserve a stretch of beach

and keep the sand from being washed away.

_____

Extensions

1. What can be done to preserve a recreational beach area from erosion as a result of excessive use by people?

Walkways can be built over dunes, protective vegetation can

be planted, and houses can be built somewhere other than directly

on the dunes.

2. What can be done to preserve a recreational beach area from being washed away as a result of wave action and longshore currents?

Beaches can be preserved with rock pilings and other jetties

built out into the water. This would deter the longshore movement

of sand. Dredging can replace beach sand lost by wave action

and longshore currents.

M O D E R N   E A R T H   S C I E N C E

Chapter 17: The Rock Record
In-Depth Investigation: Fossils

Objective
In this investigation, you will use various methods to make models of trace fossils.

Prelab Preparation
1. Describe three ways in which fossils can be classified based on how they were formed.

   (1) unaltered remains; (2) altered remains; (3) indirect evidence

2. Why are coal, oil, and gas called fossil fuels?

   Because they are the remains of ancient plants and animals.

Observations
1. **a.** Is the indentation left by the object a mold or a cast?

   A mold

   b. What features of the object are best shown in the indentation?

   Size, shape, surface markings

   c. Sketch the indentation in the space below.

2. **a.** Do the pieces of hardened plaster represent molds or casts?

   Casts

   b. Sketch your fossil imprint in the space below.

Analysis and Conclusions

1. List all objects that you can identify from the molds and casts.

Answers will vary depending on the objects available to the class.

_____

_____

_____

2. How does the carbon-print you made differ from an actual carbon-print trace fossil?

The leaf used in the investigation was given an artificial

coating of carbon. An actual print is made from the carbon left

behind from a decayed leaf.

3. Why are carbon-prints, molds, and casts trace fossils?

Because all three types of fossils are evidence of an organism,

such as its outline, its shape, or a footprint, but are not a preserved

part of the actual organism.

Extensions

1. Which of the organisms in Figure 17.1 would be most likely to form fossils? Explain.

The snail and clam, because of their hard shells, and the rabbit,

because of its bones, would readily form fossils. The fly may

be preserved in amber.

2. Which of the organisms in Figure 17.1 would leave trace fossils? Explain.

The earthworm would be most likely to leave a trace fossil in

the form of a burrow or trail in the soil. The rabbit may leave footprints.

The clam might leave a burrow and the snail a trail or track.

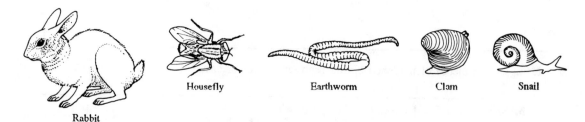

Housefly   Earthworm   Clam   Snail

Rabbit

Figure 17.1

M O D E R N   E A R T H   S C I E N C E

Chapter 18: A View of the Earth's Past
In-Depth Investigation: Geologic Time

Objective
In this investigation, you will convert the vast periods of geologic time into a frame of reference that is easier to understand.

Observations
1. Complete the following scale.

   1 meter = _____1,000,000,000_____ years

   1 centimeter = _____10,000,000_____ years

   1 millimeter = _____1,000,000_____ years

2. Convert each of the ages listed below into its equivalent distance in meters, centimeters, or millimeters.

Geologic Event	Age	
Age of the earth	4.6 billion years	4.6 meters
Oldest known plants	2 billion years	2.0 meters
Oldest known animals	1.2 billion years	1.2 meters
Paleozoic Era begins (Cambrian Period)	570 million years	0.570 meters
Ordovician Period begins	505 million years	0.505 meters
Silurian Period begins	438 million years	0.438 meters
Devonian Period begins	408 million years	0.408 meters
Mississippian Period begins	360 million years	0.360 meters
Pennsylvanian Period begins	320 million years	0.320 meters
Permian Period begins	286 million years	0.286 meters
Mesozoic Era begins (Triassic Period)	245 million years	0.245 meters
Jurassic Period begins	208 million years	0.208 meters
Cretaceous Period begins	144 million years	0.144 meters
Cenozoic Era begins (Paleocene Epoch)	65 million years	0.065 meters
Eocene Epoch begins	58 million years	0.058 meters
Oligocene Epoch begins	37 million years	0.037 meters
Miocene Epoch begins	24 million years	2.6 centimeters
Pliocene Epoch begins	5 million years	0.5 centimeters
Pleistocene Epoch begins	2 million years	2 millimeters
Holocene Epoch begins	11 thousand years	0.011 millimeters

3. Where did you place the following events on the adding-machine paper? Tell the correct age(s) and the meter conversion(s) for each event.

   a. trilobites appear, flourish, and become extinct 600–245 million years (0.600–0.245 m)

   b. first flowering plants (angiosperms) appear 135 million years (0.135 m)

   c. apes appear in Asia and Africa 26 million years (2.6 cm)

   d. dinosaurs appear, flourish, and become extinct 245–65 million years (0.245–0.065 m)

   e. first birds appear 160 million years (0.160 m)

   f. first reptiles appear 290 million years (0.290 m)

   g. first mammals appear 230 million years (0.230 m)

   h. early humans appear Answers may vary, beginning about 3 million years (3 cm)

Analysis and Conclusions

1. Explain why you cannot mark the position of important events in history such as the American Revolution on your paper scale.

   The distance to mark on the tape could not be measured for such

   a small time period.

2. Would it be possible to measure a distance on the tape that could represent the length of a person's lifetime? Explain your answer.

   No. The space would be so small you could not see it.

Extensions

1. What might be some reasons why we do not see the same animals and plants on earth today that could be found thousands of years ago?

   It cannot be determined exactly why these animals and plants died off.

   However, it can be stated that they no longer adapted to their environment

   once certain conditions prevailed.

2. Why do some animals and plants, whose origins go back millions of years, still exist today relatively unchanged except for their size?

   Although smaller in number today, they have adapted to their present

   environment just as they were able to adapt to conditions millions of

   years ago.

M O D E R N   E A R T H   S C I E N C E

Chapter 19: The History of the Continents
In-Depth Investigation: History in the Rocks

Objective
In this investigation, you will discover how the geologic history of an area can be determined from the arrangement of fossils and rock layers.

Prelab Preparation
Define the terms *index fossil* and *law of superposition* as they relate to the study of rock layers.

An index fossil is a guide fossil that can tell the relative age

of the rock in which it is found. The law of superposition states

that in undisturbed sedimentary rock a rock layer is older than the layer

above it and younger than the layer below it.

Observations
1. List the fossil names from oldest to youngest in each arrangement.

    Arrangement 1: cephalopod, blastoid, brachiopod, echinoid, pelecypod

    Arrangement 2: trilobites, brachiopod, gastropod, shark tooth,

    cephalopod (the last three fossils in Arrangement 2 may be

    listed in any order)

    Arrangement 3: trilobites, brachiopods

    Arrangement 4 trilobite, cephalopods, echinoid

2. In the four arrangements shown in Figure 19.2, do the fossils appear in the correct order of their geologic times? Name fossils, if any, that appear out of order in each arrangement.

    Arrangement 1: brachiopod is out of order.

    Arrangement 2: fossils are shown in a correct order.

    Arrangement 3: trilobites are out of order.

    Arrangement 4: fossils are shown in a correct order.

3. Do the fossils in the four arrangements represent complete sequences of geologic periods? If not, which periods are missing in each arrangement?

    Arrangement 1: missing periods are Triassic, Devonian, and Silurian.

    Arrangement 2: missing periods are Triassic, Pennsylvanian, and

    Mississippian.

    Arrangement 3: missing periods are Quaternary, Tertiary,

    Cretaceous, Triassic, and Mississippian.

    Arrangement 4: missing periods are Quaternary, Tertiary, and

    Pennsylvanian.

Analysis and Conclusions

1. What processes or events might explain the order in which each of the fossil arrangements were found?

   Answers will vary, but may include deposition, unconformities,

   folding, and faulting.

2. Based on your observations in the procedure, why is it necessary that a fossil be found over a wide geographic area in order to be considered an index fossil?

   Geologic processes such as folding and faulting change the

   arrangement of rock layers. Thus, when a certain fossil is found

   over a wide geographic area, geologists can often use it to match

   rock layers of a certain age throughout the world. If this is

   possible, the fossil may be classified as an index fossil.

3. Give a few possible reasons why there is a rock layer in Arrangement 3 containing no fossils in between rock layers that contain fossils.

   The rock layer may be very old Precambrian rock that was moved

   upward by faulting, folding, or an unconformity, or this rock layer

   may represent a span of time in which metamorphic activity destroyed

   any fossils that were present. Other answers are acceptable.

Extensions

1. Collect fossils found in your area. Identify the fossils you have collected and describe what your area was like when the organisms existed.

   Answers will vary.

2. What types of sedimentary rock usually have the most fossils?

   Limestones are the most likely sedimentary rocks to contain a

   large number of fossils. Shales are good sources of fossils

   but not all layers of shale contain fossils. Sandstones are not likely

   to contain fossils because these sediments usually accumulate on land.

3. How do index fossils aid in the discovery of oil?

   While drilling and removing core samples, petrologists look for these index

   fossils in the sample to help them determine if it would be wise to drill

   deeper for oil at that location.

M O D E R N   E A R T H   S C I E N C E

Chapter 20: The Ocean Basins
In-Depth Investigation: Ocean-Floor Sediments

Objective
In this investigation, you will determine the relationship between the size of sediment particles and their settling rate in water.

Observations
Record the time measurements in Table 20.1. Calculate and record the averages.

Table 20.1

Soil samples	Trial 1	Trial 2	Trial 3	Average
Coarse	First time measurement:			
Coarse	Second time measurement:			
Medium	First time measurement:			
Medium	Second time measurement:			
Medium-fine	First time measurement:	Answers will vary.		
Medium-fine	Second time measurement:			
Fine	First time measurement:			
Fine	Second time measurement:			

1. What is the average settling time recorded for the first particles in the coarse sample to reach the finish line? What is the average settling time for all the coarse particles to reach the finish line?
   Answers will vary.

   _____

2. What are the average settling times recorded for the medium-sized particles? The medium-fine–sized particles? The fine particles?
   Answers will vary.

   _____

3. Do layers of sediment appear in the cylinder in Step 8?
   Yes

4. Why does the water remain slightly cloudy even after most of the particles have settled in Step 8?
   Very fine particles such as clay remain suspended for long

   periods of time. Some particles are so small that they do not settle

   under the influence of gravity.

Analysis and Conclusions

1. How does the settling time of the medium particles compare with the settling time of the medium-fine particles?

It takes a slightly longer time for the medium-fine particles

to settle than it does for the medium particles.

_____

2. Do similar-sized particles fall at the same rate?

Yes

3. Other than size, what factors would you expect to influence how rapidly particles fall in sea water?

The amount of turbulence in the water, underwater landslides

or storms, and upwelling.

_____

4. How do the results in Step 8 of the procedure help to explain why the deep ocean basins are covered with a very fine layer of sediment while areas near the shore are covered with coarse sediment?

The coarse sediment settles nearer the shore because of its

heavier weight, while the fine, lightweight sediments are carried

further out to sea and settle slowly in the deep ocean basins.

Extension

What are some sources of the sediment that is found in the ocean?

Answers will vary, but may include weathered bits of rock

from the land; soil that has been washed away by erosion or

volcanic activity.

M O D E R N   E A R T H   S C I E N C E

Chapter 21: Ocean Water
In-Depth Investigation: Ocean Water Density

Objective
In this investigation, you will observe the effects of temperature and salinity on the density of salt water.

Prelab Preparation
Define density.

Density is the mass per volume of a substance or $D = m/V$.

Observations
1. Record the temperature of the salt water in Table 21.1.
2. Record the density of the water at room temperature, 25°C, and 30°C in Table 21.1.
3. Record the density of the water after 5, 10, and 15 minutes of boiling in Table 21.2.

Table 21.1

Temperature (°C)	Density (cm above or below the red line)
Answers will vary	Densest
25	
30	Least dense

Table 21.2

Minutes of boiling	Density (cm above or below the red line)
5	Least dense
10	
15	Densest

4. As the temperature increases, does the water's density increase or decrease?

   Decrease

5. Did the density of the water increase or decrease with each boiling?

   Increase

6. Originally there was 100 mL of water in the beaker. How much is in the beaker now?

   Answers will vary, but there will be less than 100 mL.

Analysis and Conclusions

1. Explain how and why warming the water affected its density.

As water warms, the water molecules move farther apart. The mass

of a given volume of the expanded water, and thus the density,

is less.

2. Based on your observations, infer what the density of polar ocean water would be compared with the density of water of equal salinity near the equator. Explain your answer.

The polar water would be denser because it receives less infrared

radiation than the water near the equator.

3. Why did the amount of water in the beaker change? Explain why boiling the water affected its density.

While water boils, its temperature remains the same. The amount

of water in the beaker decreased because some water evaporated. The

density of the water increased because, while the amount

of water decreased, the amount of salt in the water did not. Thus,

its salinity, and therefore its density, increased.

Extension

Place a beaker of salt water in a freezer until a crust of ice forms at the top. Break up and remove the ice from the water. Is the water denser or less dense than before freezing? Explain.

The water is denser than before freezing. Temperature and salt

affect the density of water. Water becomes more dense since fresh water

is removed with the ice, leaving salt in a lesser amount of water.

M O D E R N   E A R T H   S C I E N C E

Chapter 22: Movements of the Ocean
In-Depth Investigation: Wave Motion

Objective
In this investigation, you will simulate wave motion in order to observe how energy generates wave motion on water. You will also observe various properties of waves.

Prelab Preparation
1. Label the wave crests and wave troughs on Graph 22.1.

Graph 22.1

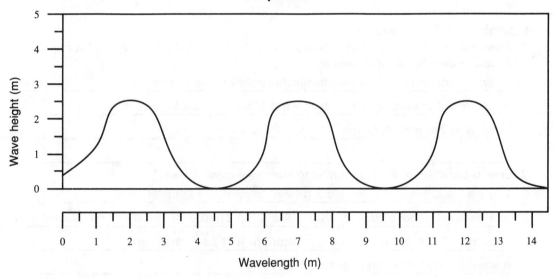

2. **a.** What is the wavelength in Graph 22.1?

    5 m
    _____

    b. What is the wave height?

    2.5 m
    _____

    c. What is the speed of the waves in Graph 22.1 if the wave period is five seconds?

    5 m ÷ 5 sec = 1 m/sec
    _____

Observations
1. Plot the waves you simulated with the rope on Graph 22.2, page 104.

2. **a.** What are the wavelengths of the three waves you plotted?

    Answers will vary.
    _____

    b. What are the wave heights of the three waves?

    Answers will vary.
    _____

    c. What are the wave speeds of the three waves if each wave period is six seconds?

    Answers will vary.
    _____

Graph 22.2

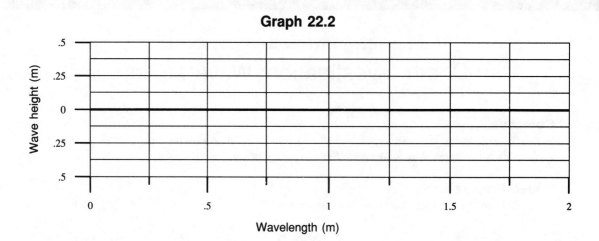

Analysis and Conclusions

1. How do the wave motions differ on your graph? If these were real water waves, what might be the cause(s) of the different motions?

   Answers will vary. Students should note that wave motion in

   water varies according to certain wind factors, such as intensity,

   fetch, and the length of time the wind blows.

2. How is the action of the rope similar to wave movement in water?

   The waves created on the rope moved toward the secured end,

   just as waves move toward shore. Also, the energy of the hand movement

   determined the speed of the rope motion, just as energy from

   the wind determines the speed of waves.

3. How do the motions of the cloth ties differ from the wave's motion?

   The ties move side to side while the waves move along the rope

   toward the chair or table leg.

4. What do the motions of the ties tell you about wave motion in water?

   As waves pass, the ties move side to side with little forward

   movement. These motions show that energy creates wave motion in water,

   and that water actually moves very little.

Extension

Describe the five waves you plotted on your graph. Compare the waves plotted using a 4-m rope to the waves plotted using a 2-m rope. Are there differences and/or similarities? Explain. What are the wave speeds that you calculated for each wave?

Answers will vary. Students should note that the extra length

in the rope enables them to create more varied types of waves in terms

of speed and height.

M O D E R N   E A R T H   S C I E N C E

Chapter 23: The Atmosphere
In-Depth Investigation: Air Density and Temperature

Objective
In this investigation, you will observe how the density of air is affected by changes in its temperature.

Prelab Preparation
1. A block of lead has a mass of 910 g and a volume of 80 cm^3. What is lead's density?
   About 11.37 g/cm^3
2. If the mass of a given substance remains the same while its volume increases, will the density of the substance increase or decrease?
   decrease

Observations
1. Record your observations in Table 23.1.

Table 23.1

Temperature (°C)	Volume (cm^3)		
	Pull	Push	Average
	Answers will vary		

2. Plot your data in Graph 23.1 on page 108.
3. List reasons why the points in Graph 23.1 do not all fall on a straight line.
   Sticking plunger, inaccurate measurements

Analysis and Conclusions
1. According to your graph, what is the increase in volume, in cubic centimeters, as the temperature increases from 20°C to 40°C.
   Answers will vary.
2. As the temperature increases, does the density of the air in the syringe increase or decrease? Explain.
   Decrease. As the temperature increases, the volume increases.
   If the mass remains the same, the density decreases (D = m/V).

Graph 23.1

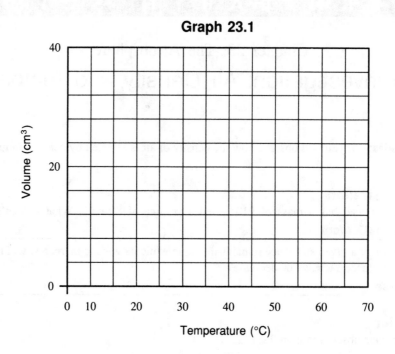

3. Is hot air less dense or more dense than cold air?
   Less dense
4. If air is warmed, does it become less dense or more dense?
   Less dense
5. Would warm air tend to rise or sink in a body of colder air?
   Rise

Extensions

1. How might a temperature inversion affect air quality?
   Polluted air is usually replaced by fresh air when convection

   occurs. When the air cannot move, pollutants build up, creating

   unhealthy breathing conditions.
2. What might cause conditions to return to normal?
   Wind might blow pollutants from the area. The cool air

   near the ground might warm up by absorbing heat from the sun and rise.

M O D E R N   E A R T H   S C I E N C E

Chapter 24: Water in the Atmosphere

In-Depth Investigation: Relative Humidity

Objective
In this investigation, you will use wet-bulb and dry-bulb thermometer readings to determine relative humidity.

Prelab Preparation
1. Study Graph 24.1.
   a. How many grams of water can a cubic meter of air hold at a temperature of 20°C?

   about 17 g _____

   b. How many grams can the air hold at 40°C?

   about 53 g _____

Graph 24.1

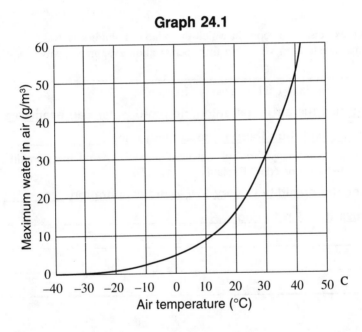

2. If a cubic meter of air at 40°C holds 40 g of water, what is the relative humidity?

   40 g/53 g = 0.75 = 75% _____

Observations
1. **a.** Will the two thermometers have the same reading?

   No. _____

   b. If not, which thermometer will have the lower reading?

   The wet-bulb thermometer. _____

2. a. What is the temperature on the dry-bulb thermometer?

Answers will vary.

b. What is the temperature on the wet-bulb thermometer?

Answers will vary.

c. What is the difference in the two temperature readings?

Answers will vary.

3. What is the relative humidity at your laboratory station? Use Table 24.1 on page 111.

Answers will vary.

Analysis and Conclusions

1. Based on the relative humidity you found, can the air in your classroom hold more evaporated water?

Yes, unless the relative humidity is 100%.

2. If you wet the back of your hand, would the water evaporate and cool your skin?

Yes, unless the relative humidity is 100%.

Extensions

1. Suppose you exercise in a room in which the relative humidity is 100%.
a. Would the moisture on your skin from perspiration evaporate easily?

No

b. Would you be able to cool off readily? Explain.

No. Because the air is saturated, no more water can evaporate

into the air and thus there is no cooling effect.

2. Suppose you have just stepped out of a swimming pool. The relative humidity is 30%. How would you feel—warm or cool? Explain.

Cool. Since the humidity is very low, rapid evaporation

would cause the skin to cool.

M O D E R N   E A R T H   S C I E N C E

Chapter 25: Weather

In-Depth Investigation: Weather Map Interpretation

Objective

In this investigation, you will study the symbols used on a weather map to gain an understanding of the relationship among temperature, pressure, precipitation, and winds.

Observations

1. On the map shown in Figure 25.1, draw the isotherms in red for every 2°C of temperature beginning with the 10°C isotherm.
2. Next, draw the isobars in blue for every 4 millibars of pressure, beginning with the 1004-millibar isobar.

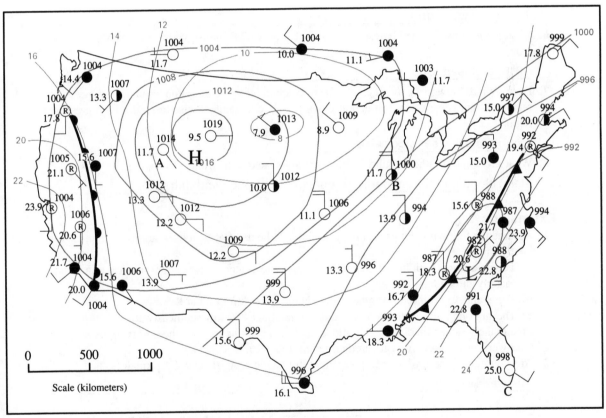

Figure 25.1

Analysis and Conclusions

1. What is the lowest temperature for which you have drawn an isotherm? What is the highest temperature for which you have drawn an isotherm?

8°C; 24°C

2. Is either isotherm a closed loop? If so, which one?

Yes, the lowest

3. Is the air mass identified by the closed isotherms a cold air mass or a warm air mass?

Cold air mass

4. Is there a shift in wind direction associated with either front shown on your map? Describe the shift.

Yes; the winds ahead of the cold front shift from a south/southeasterly direction to

a northerly direction as the front passes. The winds ahead of the warm front show

little or no shift from their south/southeasterly direction as the front moves east.

5. What is the value of the lowest-pressure isobar drawn? Is the isobar an open curve or a closed loop?

984; a closed loop

6. What is the value for the highest-pressure isobar drawn? Is the isobar an open curve or closed loop?

1016; a closed loop

7. Do the winds blow in a general clockwise or counterclockwise direction around a low-pressure region? Is the general wind direction clockwise or counterclockwise around a high-pressure region?

Counterclockwise; clockwise

8. What are some weather conditions that are associated with fronts?

Rain, clouds, rapid temperature change, dramatic windshift

Extensions

1. Assume the following things will happen in the 24 hours after the observations were made for your map.
  a. The cold front, with its weather and low-pressure center, moves about 650 km in a southeasterly direction.
  b. The warm front and its weather move about 1000 km eastward.
  c. The high-pressure center and weather move about 800 km southeastward.
  Predict the weather conditions at Stations *A, B,* and *C* 24 hours after the observations for your map were made. Record your predictions in Table 25.1.

2. Are you equally sure of all the predictions for Stations *A, B,* and *C*? Explain.

Answers will vary.

Table 25.1

Station	Pressure	Wind direction	Wind speed	Temperature	Sky condition
A					
B					
C					

MODERN EARTH SCIENCE

Chapter 26: Climate

In-Depth Investigation: Factors That Affect Climate

Objective

In this investigation, you will explore how the angle of the sun's rays and the distribution of land and water affect climate.

Observations

1. How many squares on the graph paper were lit in Step 4? How many were lit in Step 5?

   Answers will vary, but the angled beam will illuminate more squares.

2. What area(s) of the earth is represented by the light striking the graph paper at a 90° angle? What area(s) of the earth is represented by the 45° angle?

   Answers will vary.

3. What was the recorded temperature of the water after 3 minutes? After 10 minutes? Five minutes after the light was turned off?

   Answers will vary, but the water will heat and cool more slowly than the soil.

Analysis and Conclusions

1. Compare the area of the lighted squares obtained from the two trials. Which area is greater?

   The area that is on the 45° angle is greater.

2. What area(s) of the earth is represented by the flashlight striking the graph paper at a 90° angle? What area(s) of the earth is represented by the 45° angle?

   At 90°, near the equator; at 45°, light strikes areas north

   and south of the equator in middle-latitude climates.

3. Why do the sun's rays from overhead cause more warming than low-angle rays?

   When the sun's rays strike the earth directly, the surface receives

   more energy per unit area than when the rays strike at an angle.

4. What conclusion can you draw about the angle of the sun's rays during the different seasons in the United States?

   During the summer, the sun's rays strike the United States directly.

   During the winter, the sun's rays strike the United States at an angle. This

   creates warmer temperatures in the summer and colder temperatures in the

   winter.

5. Which substance, water or soil, absorbed more heat in Step 9?

   The soil

6. Which substance, water or soil, lost heat faster when the heat source was turned off in Step 10?

   The soil

7. What conclusion can you draw about how land and water on the earth are heated by the sun?

Land heats faster and to a higher temperature than water does.

Water, however, releases heat more slowly than land does so it stays

warmer for longer periods of time.

8. Explain how the light striking the graph paper in Steps 4 and 5 is related to the effect of heat absorption in water and soil on the earth, as simulated in Steps 9 and 10.

The way the sun's rays strike the earth and are absorbed or

reflected affects the surface temperatures on the earth. Thus, an

area's climate is determined by its distribution of water and

land and the angle at which the sun's rays heat the water and land.

Extensions

1. Continue this investigation using the flashlight and graph paper. Hold the flashlight at various angles to the paper such as at a 20° angle, 0° angle, and so on. What areas of the earth are represented on the paper at the different angles you use?

Answers will vary. Students should note that as the sun's rays

spread out further, such as at a 20° angle, they are less intense per unit area,

creating colder climates in these areas of the earth.

2. Considering the earth's varied climates, what conclusions can you draw concerning the angle of the sun's rays on different locations on the earth?

At the equator the sun's rays hit the earth at almost 90°.

Therefore, the temperatures are high. As you go north or south from

the equator the sun's rays hit the earth at varying angles,

and temperatures decrease as you go in either direction.

M O D E R N   E A R T H   S C I E N C E

Chapter 27: Stars and Galaxies
In-Depth Investigation: Star Magnitudes

Objective
In this investigation, you will determine the effect of distance on brightness and the relation between temperature and color.

Observations
1. What are the distances from each bulb to the center of the photometer?

   Answers will vary, but distances should be approximately 42 cm from the

   bulb with one battery and 58 cm from the bulb with two batteries.

2. Which way did you move the photometer to make it record equal brightnesses on both sides?

   Toward the fainter bulb

3. Square the distances you recorded in Step 1 of the Observations.

   $42^2 = 1764$; $58^2 = 3364$

4. Compare the color differences between the paraffin sides of your photometer.
   a. Is the light bulb or sunlight more yellow?

   The light bulb

   b. Is the light bulb or sunlight more white?

   The sunlight

5. Compare the colors of the bulb powered by a single battery and the bulb powered by two batteries in the darkened room. What did you observe?

   The single-battery bulb is more yellow, while the double-battery bulb is

   whiter.

Analysis and Conclusions
1. As you move away from a light source, the brightness decreases in relation to the square of the distance. The ratio of the square of the distances you calculated in Step 3 of the Observations is equal to the ratio of the brightnesses of the bulbs. What is the ratio of the square of the distance of the two-battery flashlight to that of the one-battery flashlight? What does this tell you about the relation of the brightnesses of the two flashlights?

   3364/1764 = 1.9. Because the ratio of the squares of the distance

   is about 2 to 1 then the two-battery bulb must be about twice as bright

   as the one-battery bulb. Students' answers, however, will depend

   on their results.

2. An astronomer knows that two stars have the same spectra and should be of the same brightness. Yet one is four times fainter than the other. How much further away is the fainter star?

   The fainter star is twice as far away.

3. Based on the results of the investigation, would you expect a white star or a yellow star to be hotter?

A white star

4. Would you expect a white star to be hotter or cooler than an orange star? Predict whether a blue star is hotter or cooler than a white star. Predict whether a red star is hotter or cooler than an orange star.

A white star is hotter than an orange star; a blue star is hotter

than a white star; a red star is cooler than an orange star.

Extensions

1. Find an incandescent bulb controlled by a dimmer. Watch the color of the light as it fades. Does it get more yellow or more white? Explain why.

It gets more yellow. As the dimmer is turned down, the light

becomes less intense. The light fades to a yellow color because it is cooler.

2. If a red star is very cool with a low absolute magnitude, why are bright red stars visible in the sky?

The bright red stars must be very large or very close to us.

M O D E R N   E A R T H   S C I E N C E

Chapter 28: The Sun

In-Depth Investigation: Size and Energy of the Sun

Objective

In this investigation, you will perform a simple experiment from which you can calculate the sun's diameter. Then you will collect energy from sunlight and estimate the amount of energy produced by the sun.

Prelab Preparation

Describe how a pinhole image of the sun is formed.

Sunlight passes through a small hole and a tiny image is formed behind

the pinhole on the card.

Part I Observations

1. What is the diameter of the sun's image?

   Answers will vary

2. What is the distance between the image and the pinhole?

   Answers will vary

3. The ratio of the diameter of the sun's image to its distance to the pinhole is equal to the ratio of the sun's diameter divided by the distance to the sun as shown by the equation:

$$\frac{\text{sun's diameter}}{\text{sun's distance to pinhole}} = \frac{\text{image diameter}}{\text{image's distance to the pinhole}}$$

If the distance between the sun and the earth is 150 million kilometers, calculate the diameter of the sun using your data.

Should be about 1.3 million kilometers, though answers may vary widely

Analysis and Conclusions

1. What is the sun's diameter as given in the textbook?

   1,300,000 km

2. Explain any difference in your answer and the textbook value.

   Difficulties in making exact measurements; box didn't lie flat,

   thus changing the angle of the sun's rays.

Part II Observations

1. What is the temperature reading on the thermometer when the solar collector faces the sun?

   Answers will vary

2. What is the distance between the center of the lamp and the thermometer bulb?

   Answers will vary

Analysis and Conclusions

1. The collector absorbed as much energy from the sun at a distance of 150 million km as it did from the 100-W bulb at the distance you measured. Therefore, the following equation can be used.

$$\frac{\text{power of the sun (in watts)}}{(\text{distance to sun})^2} = \frac{\text{power of the lamp (in watts)}}{(\text{distance to lamp})^2}$$

The distance to the sun is 1.5×10^{13} centimeters. Use the equation above to calculate the power of the sun. Be sure that you express both distances in the same units.

Should be about 3.7×10^{26} watts, though answers will vary.

2. The sun's power is quoted as 3.7×10^{26} watts. Compare your answer with this value. Describe any sources of measuring inaccuracy with your method. Do not be surprised if your answer is 100 times too large or small. This is a very crude method to approximate a very large number.

Sources of error include heat from light in the room; heat exchange

between jar and room, difference in temperature of the sun and the light

bulb.

Extension

Would the experiment have worked with a flourescent bulb? Explain.

Not as well, because fluorescent bulbs produce light in a different

manner and for much less energy.

Name _____ Class _____ Date _____

Chapter 29: The Solar System
In-Depth Investigation: Crater Analysis

Objective
In this investigation, you will experiment with making craters to discover the effect of speed and projectile angle on the crater formed.

Observations
Write a description of each of the craters and the surrounding area.

Crater *A*   A circular crater with moderately high walls. Some material has splashed out around the crater.

Crater *B*   A deep circular crater with high walls. A large amount of material is splashed out around the crater

Crater *C*   A shallow circular crater with low walls. Very little or no visible material ejected from crater.

Crater *D*   Very large, deep crater with very high walls. Great amount of material splashed out around crater. Material splashed far from the center of the crater.

Crater *E*   A circular crater of moderate depth, one wall higher than the other. Material splashed out of crater is more on the low-walled side of crater.

Crater *F*   Same description as crater *E*.

Analysis and Conclusions
1. Which crater was formed by the marble with the highest velocity? What is the effect of velocity on the characteristics of the crater formed?
   B; a higher velocity produces a larger crater with more material ejected around the crater.

2. Study the shapes of craters *A, E,* and *F.* How does the angle at which the marble strikes the surface affect the shape of the craters formed?

   All three craters are circular. However, the walls of craters *E* and *F*

   are higher on the side of the crater toward which the marble was traveling.

   There is no visible difference between craters *E* and *F.*

3. Compare craters *A, B,* and *D.* How do they differ? What caused this difference? Is the difference in the masses of the objects a factor? Explain.

   Craters *B* and *D* are deeper and have larger diameters. Also more material

   is ejected from craters *B* and *D* and the material is ejected farther from the

   craters than is the material ejected from crater *A;* crater *D* was formed by a

   larger object; the difference in mass is not a factor because objects fall at

   the same velocity no matter what the object's mass is; crater *B* formed a

   larger crater due to its greater speed.

Extension

Which type of crater formation have we not studied? Devise a method of studying this cratering process using the model you made.

Volcanic craters; either spread plaster of Paris over chicken wire or poke

holes in the bottom of a box filled with the plaster, then blow air

upward from underneath.

M O D E R N   E A R T H   S C I E N C E

Chapter 30: Moons and Rings
In-Depth Investigation: Galilean Moons of Jupiter

Objective
In this investigation, you will verify that the orbital motion of Jupiter's moons obey Kepler's third law.

Prelab Preparation
Find out what is meant by a constant and a variable.

A constant is a quantity in an equation that always has the same value.

A variable is a quantity with no fixed value.

Procedure
1. **a.** Using Figure 30.2 on page 140, list the days when each of Jupiter's moons cross in front of the planet.

   Io: 1, 3, 5, 6, 8, 10, 12, 13, 15, 17, 19, 20, 22, 24, 26, 28, 29, 31; Europa:

   0, 3, 7, 10, 14, 17, 21, 24, 28, 31; Ganymede: 6, 13, 20, 27; Callisto: 6, 23

   b. List the days when the moon's are behind the planet.

   Io: 0, 2, 4, 5, 7, 9, 11, 12, 14, 16, 18, 20, 21, 23, 25, 27, 28, 30; Europa: 1, 5,

   8, 12, 16, 19, 23, 26, 30; Ganymede: 2, 9, 16, 23, 31; Callisto: 14, 31

2. Use the data in Table 30.1 to test Kepler's third law. Calculate p^2 and r^3 for each of the planets. Record your results in the table. Then calculate K for each planet using Kepler's third law, $K = p^2/r^3$. Record your results in the table.

Table 30.1 The Planets

Planet	r (in billions of kilometers)	p (in Earth years)	r^3	p^2	K
Mercury	0.058	0.24	0.000195	0.058	297
Venus	0.108	0.62	0.00126	0.38	302
Earth	0.150	1	0.00338	1	296
Mars	0.228	1.88	0.01185	3.53	298
Jupiter	0.778	11.86	0.47091	140.7	299
Saturn	1.427	29.46	2.9058	867.9	299
Uranus	2.869	84.01	23.615	7,058	299
Neptune	4.486	164.8	90.277	27,159	301
Pluto	5.890	247.7	204.34	61,355	300

3. According to your results in Table 30.1, is K a constant?

   Yes, about 299.

4. Draw Jupiter and its moons as they would appear from the earth at midnight on the 2nd and 26th of the month in the eyepiece of the telescope shown in Figure 30.3 on page 144.

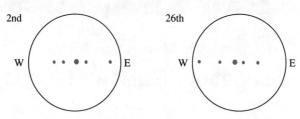

Figure 30.3

5. In Figure 30.4, draw Jupiter's moons on the first day of the month that all four moons are on the same side of the planet. Give the date.

6. Give a date when only two moons will be visible. Name the two visible moons.

   <u>3, 10, Ganymede and Callisto; 14, Europa</u>

   <u>and Ganymede; 23, 30, Ganymede and</u>

   <u>Callisto; 31, Io and Europa</u>

7. In Table 30.2, record the length of time in earth days required by each moon to orbit Jupiter.

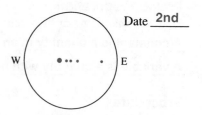

Date <u>2nd</u>

Figure 30.4

Table 30.2 Galilean Moons

Moon	p (in Earth days)	Scale r (in cm)	p^2	r^3	K
Io	1.8	0.25	3.24	0.016	202.5
Europa	3.6	0.4	12.96	0.064	202.5
Ganymede	7.4	0.65	54.76	0.275	199.1
Callisto	16.5	1.1	272.25	1.331	204.5

8. In Table 30.2, record the scale distance between the maximum outward swing of each moon and the center of Jupiter in centimeters.

9. Square each period measurement and record the answer in Table 30.2. Cube each distance measurement and record the answer in the table.

10. Use your results to test Kepler's third law. Because $K = p^2/r^3$, divide p^2 by r^3 for each moon to find K. Record your results in Table 30.2.

Analysis and Conclusions

1. Will you see all four of Jupiter's largest moons each time you look at Jupiter through a telescope or binoculars? Explain.

   <u>No, sometimes a moon will be behind or in front of Jupiter.</u>

2. Jupiter's moons look like dots in a telescope. You cannot tell them apart by their appearance. If you had no charts, how could you identify each moon?

   <u>By observing their maximum outward swing and by determining the periods.</u>

3. After you solve for K for each of the moons, study your results. Is K a constant?

   <u>Yes, the numbers all about equal.</u>

Teacher's Guide to Long-Range Investigations

About Long-Range Investigations

The ten *Long-Range Investigations* found both in the back of the textbook as well as in a consumable booklet involve students working in astronomy, hydrology, oceanography, meteorology, and climatology. Ideally, the long-range investigations should demonstrate the relevance of earth science to the student, the school, and the community.

Long-range investigations can be assigned prior to, during, or after teaching a related chapter or unit. The observation schedules for these investigations range from about one week to several months. Where the time frame is longer, the intervals between observations generally are longer. Therefore, student effort is roughly equal for each investigation.

Several of the long-range investigations—Positions of the Sunrise and Sunset, Comparing Climate Features, and Planetary Motions—should be assigned at the beginning of the academic year. These three investigations have extended observations periods. As the topics of these investigations become pertinent to classroom lessons, the selected students could report on their observations. The entire class could contribute to the analyses and conclusions for the investigation. Throughout the year, you may want to schedule time for periodic progress reports from the students who have been assigned these long-range investigations.

The *Teacher's Guide and Answer Keys to Investigations* contains teacher notes and annotated lab report pages for each long-range investigation. The long-range investigations are not arranged to follow the sequence of each chapter. However, chapter correlations and textbook page numbers are provided in the notes for each investigation. Thought-provoking questions and extension activities are suggested in the "For Further Investigation" section of the teacher strategies. These questions extend the concepts covered in the investigation. Many of the questions are open-ended and can be used to guide students to expand their critical thinking and science skills.

Contents for Long-Range Investigations

Long-Range Investigation 1
Positions of Sunrise and Sunset

Purpose
To investigate the sun's change of position along the horizon at sunrise and sunset.

Skills
analyzing, comparing, inferring, interpreting data, measuring, observing

Planning
Students will need approximately 8 months to complete this investigation. Have the students begin this investigation in September. You can have students present their data and conclusions after they complete their investigations in May (at the end of the school year).

Student-Textbook Correlation
Descriptions of the seasonal changes in the angle and duration of daylight are found in the textbook in Chapter 2, Section 2.2 Movements of the Earth, pages 29–32.

Cooperative Learning
1. Have students share the responsibilities for observing sunrise and sunset directions. Early risers should be matched with students who find it difficult to get up before dawn.
2. Assigning students to staggered dates within each month will provide for nearly continuous observations.

TEACHING STRATEGIES
Advance Preparation
1. Tongue depressors may be obtained from the school nurse for use in this investigation or ordered from a science-equipment supply house.
2. You may want to ask the Industrial Arts teacher to drill the holes in the tongue depressors.
3. Prepare a list of open areas or high areas in your community, where the view of sunrise and sunset would not be blocked.
4. Prepare a model graph to explain extrapolation to your students. To prepare this model use the Sample Data Graph on page T-8. Expand it to include July and August. Ask students to hypothesize the position of the sun in July and August from the information already given in the graph.

Safety Guidelines
In this investigation the following safety guidelines should be emphasized.

CAUTION: If you are using hand or power tools, it is best to allow your parents or some other adult skilled in the tool's operation to help you.

CAUTION: Although sunlight is less intense at sunrise and sunset, you should never look at the sun even for short periods of time.

Possible Problems

1. Many students will not be willing to get up before dawn to prepare for observations. See Cooperative Learning suggestions.
2. Constant alignment of the bearing chart is critical. Students who are careless in the setup will find no pattern in the graphs.
3. The direction of magnetic north usually differs from true north by a few degrees. See text page 45, Figure 3.4, for a map of magnetic declination throughout the United States. Have students adjust the alignment of the bearing chart to point true north so their graphs are accurate.

Teaching Suggestions

1. A large magnetic compass marked from 0 to 360° can be used instead of the bearing chart. A small compass is too inaccurate for the range of directions in this investigation. The compass can be used like the bearing chart.
2. Have students correct for the local declination. Magnetic declination is discussed in Chapter 3, Section 3.1 Finding Locations on the earth, page 45.
3. Emphasize to the students that proper alignment of the bearing chart is essential.

Sample Data Graph

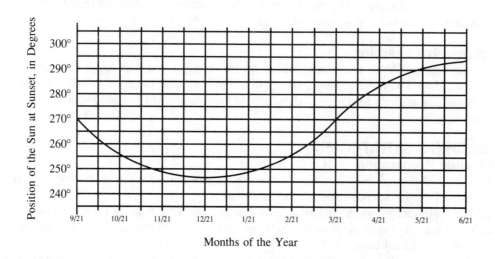

Months of the Year

For Further Investigation

1. The earth is closer to the sun during winter in the Northern Hemisphere. Have students find out why in the Northern Hemisphere it is hotter during the summer when the earth is at its furthest distance from the sun.
2. The sun reaches its highest position in the sky each day at noon. In the United States the sun reaches its zenith only in Hawaii. Why?
3. The times of sunrise and sunset vary each day. Have students investigate why this variation takes place.
4. Have students extrapolate their graphs to complete the annual cycle of the sun. Tell them to refer back to the explanation of extrapolation you gave them before they began this investigation.

M O D E R N   E A R T H   S C I E N C E

LAB REPORT

Long-Range Investigation 1
Directions of Sunrise and Sunset

Purpose

In this investigation you will examine the changing position of the sun along the horizon at sunrise and sunset.

Observations

Table 1.1 Position of the Sun

Date	Position of Sunrise in Degrees	Position of Sunset in Degrees
Sept. 21	90	270
Oct. 21	78	258
Nov. 21	70	250
Dec. 21	67	247
Jan. 21	70	250
Feb. 21	78	258
Mar. 21	90	270
Apr. 21	102	282
May 21	111	291

Analysis and Conclusions

1. On which date does the sun (a) rise the greatest number of degrees north of east? (b) set the greatest number of degrees north of west?

   Because students have only worked 9 months, they will probably answer

   May 21; the correct answer is June 21. Make sure that they are aware of

   the correct answer.

2. On which date does the sun (a) rise the greatest number of degrees south of east? (b) set the greatest number of degrees south of west?

The sun rises the greatest number of degrees south of east on about

December 21.

3. On which date(s) does the sun (a) rise approximately 90° due east? (b) set approximately 270° due west?

The sun rises due east (90°) and sets due west (270°) on about March 21

and September 22, respectively.

4. In which season is (a) the sunrise north of east? (b) the sunset north of west?

The sunrise is north of east and the sunset is north of west during the

spring and summer seasons, respectively.

5. What general statement can be made about the pattern of sunrise and sunset as shown by the graphs?

The graphs follow a curve with the highest angle about June 21 and the

lowest angle about December 21. There is a gradual curve upward from

December 21 to June 21 and a gradual curve downward from June 21 to

December 21.

6. What causes the change in the position of the sun as viewed from the earth?

As the earth moves around the sun in its orbit, the sun appears to move.

However, it is the earth that moves.

Long-Range Investigation 2
Water Clarity

Purpose
To investigate how changes in water clarity relate to changes in the environment.

Skills
analyzing, calculating, comparing, constructing, drawing inferences, hypothesizing, making and interpreting graphs

Student-Textbook Correlation
Description of water turbidity is given in the textbook, Chapter 13, Section 13.2 River Systems and 13.3 Stream Deposition, pages 247–255.

Planning
Students will need approximately one month to complete this investigation. If you plan to have students share their work, data, and conclusions with the class, be sure the investigation is completed before you begin Chapter 22.

Cooperative Learning
1. On a lake or pond, different groups may make observations at different stations. See Suggestion 1 in For Further Investigation.
2. Students could be assigned to different days to produce a more extensive data file.

TEACHING STRATEGIES
Advance Preparation
1. Although homemade Secchi discs are quite satisfactory, prefinished discs may be ordered from science-equipment suppliers.
2. Contact the industrial arts teacher in your building and explain what your students will be doing. Ask this teacher if your students could come for help when they construct their Secchi discs. Details of construction are provided in the student instructions.
3. Make a list of local bodies of water that could be used by the students. Check out the locations to be sure that students can conduct the investigation safely at each location.

Safety Guidelines
In this investigation, the following safety guidelines should be emphasized.

 CAUTION: Tell students that if they are using hand or power tools, it is best to allow their parents or some other adult skilled in the tool's operation to help them.

 CAUTION: Emphasize to the students not to go to the dock alone and that they must wear a life jacket while working near water.

Possible Problems
1. Students may select a body of water so shallow that the disc may not fade from view.
2. Most students will need prodding to suggest hypotheses for changes in turbidity. Refer to Teaching Suggestion 5.
3. Remind students to keep to their schedule of observations.

Teaching Suggestions

1. Tie in construction of Secchi disc to industrial arts. Have students work with the industrial arts teacher on the construction of their discs.
2. Washers, nuts, or any other metal objects can replace the lead sinkers. It is not necessary for the weights to be exactly 200 grams.
3. For this long-range investigation, students should work in small groups.
4. Ask several groups of students to paint their Secchi discs different colors, such as yellow and blue or gray and red. After the work is completed have these students share their results with the other groups. Does the color of the disc affect its visibility? Is a bright yellow-and-blue disc more visible or less visible than a black-and-white one?
5. Encourage the students to hypothesize about some factor that may cause the water clarity to change and then help them devise a plan to measure that factor while they are measuring the water clarity. Your knowledge of local or seasonal influences on the selected body of water will be extremely useful in planning this investigation.
6. You may wish to use this long-range investigation as a class activity and conduct a field trip to a nearby body of water.

For Further Investigation

1. You may want students to make a comparison among different locations by taking the measurements at several locations on the same body of water. Have students hypothesize what factors may have affected any variation that was found.
2. You may want to have students hypothesize why the clarity of the water in a river, lake, reservoir, or bay may be affected by storms. Make daily turbidity determinations before and after heavy rainstorms.
3. If students are making their observations in different bodies of water, such as a lake and a river, you may want to have them compare the turbidities of each. Have students hypothesize on the causes of any differences they observe.

M O D E R N   E A R T H   S C I E N C E

LAB REPORT

Long-Range Investigation 2
Water Clarity

Purpose
In this investigation, you will discover how changes in water clarity relate to changes in the environment.

Advance Preparation
1. Write the following depths in meters.

   4 m 40 cm __4.4 m__   4 m 15 cm __4.15 m__

   2 m 0 cm __2.0 m__   0 m 80 cm __0.8 m__

2. Calculate the average depth for the following sample trials.

   a. Trial 1   4.6 m   b. Trial 1   2.7 m
      Trial 2   4.4 m   Trial 2   2.9 m
      Trial 3   4.7 m   Trial 3   2.9 m

   _____4.56 m_____   _____2.83 m_____

Observations

Table 2.1 Visibility depth (m)

Location where investigation took place _____

Date	1st trial	2nd trial	3rd trial	average
9/7	3.3	3.0	3.4	3.23
9/10	3.3	3.5	3.2	3.33
9/13	4.2	4.4	4.2	4.26
9/15	4.6	4.8	4.8	4.73
9/17	3.1	3.0	3.4	3.16
9/20	2.9	2.7	2.9	2.83
9/23	2.9	2.8	2.8	2.83
9/25	3.3	3.1	3.1	3.16
9/28	3.6	3.4	3.6	3.53
10/1	3.5	3.7	3.4	3.53
10/5	4.1	4.0	4.1	4.06
10/7	4.3	4.3	4.2	4.26

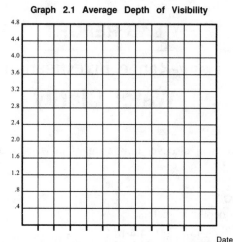

Depth in meters

Graph 2.1 Average Depth of Visibility

4.8
4.4
4.0
3.6
3.2
2.8
2.4
2.0
1.6
1.2
.8
.4

Date

Analysis and Conclusions

1. Does the graph rise and fall regularly?

Answers will vary depending upon local or seasonal conditions.

2. Does there seem to be a pattern to your graph?

Answers will vary, but cyclical patterns have relatively constant periods.

_____

3. If your graph shows a pattern, what is the amount of time from one high or low point to the next high or low point?

Answers will vary depending on the answer to question 2.

4. Do factors such as rain, wind, or tides affect water clarity? Hypothesize why these factors may change the turbidity. Explain your answer.

Answers will vary.

_____

5. As the depth at which the disc disappears increases, what can be stated about the amount of suspended particles in the water?

The number of suspended particles decreases.

_____

6. On which day was the water turbidity the highest? Hypothesize why the water turbidity was so high. Explain your answer.

Answers will vary.

_____

7. What conclusion can you make about the water clarity at your site?

Answers will vary.

_____

_____

Long-Range Investigation 3
Tides at the Shoreline

Purpose
To observe how the phases of the moon affect the tidal range.

Skills
analyzing, calculating, comparing, drawing inferences, making and interpreting graphs, measuring

Student-Textbook Correlation
Tides are explained in the textbook in Chapter 22, Section 22.3 Tides, pages 439–443.

Planning
If you wish to have students share their data and conclusions with the class, this investigation must be completed before you begin Chapter 22. This investigation will take two months.

Cooperative Learning
1. If adequate precautions are taken to minimize differences in measurement techniques, the task of making observations may be shared by students to provide a longer continuous record.
2. The tides in a bay may be observed by teams assigned to several different stations to see whether they are synchronous. See Suggestion 1 in For Further Investigation.

TEACHING STRATEGIES
Advance Preparation
1. Students should mark 1-m intervals on strings.
2. Provide students with a list of possible local locations at which to perform this investigation safely.
3. Help students locate a calendar that shows the dates of different phases of the moon.

Safety Guidelines
In this investigation the following safety guidelines should be emphasized:

CAUTION: You should not go to the water alone. Always work with a partner or an adult. Wear a life jacket while working near the water.

CAUTION: Do not make observations in any type of stormy weather.

Possible Problems
1. Students must be diligent in keeping to schedule in order to observe low and high tides.
2. Stormy weather conditions may prevent observations from being made on the proper days. Storms also cause high waves that will make taking measurements difficult.

Teaching Suggestions
1. This investigation should be completed during the spring, when the weather is fairly warm. However, it can be done at any time during the year.
2. For this long-range investigation, students should work in small groups.
3. Show a model of the earth, the moon, and the sun as they are positioned in space for the four phases of the moon used in this investigation.

4. Explain to students that during the new moon the combined gravity of the moon and the sun pull the ocean out from the solid earth. This results in a spring tide on the lighted side of the earth. The dark side simultaneously experiences spring tide. During the full moon, the moon and the sun are on opposite sides of the earth. The oceans on both sides of the earth are pulled out from the solid earth and spring tides result. During the first and last quarter phases, the line from the earth to the moon is perpendicular to the line from the earth to the sun. As a consequence, the two gravitational influences offset each other, resulting in neap tides.

5. Washers, nuts, or any other metal objects can replace the lead sinker. It is not necessary for the weight to be exactly 200 grams.

6. Tide tables may be obtained for the class by writing to:
   The National Oceanic and Atmospheric Administration
   1305 East West Highway, Silver Spring, MD 20910

7. Explain to students that the tidal bulges caused by the gravitational pull of the sun and moon remain stationary as the earth rotates. An object located on the surface of the water would be carried alternately into deeper and shallower water as the earth rotates.

For Further Investigation

1. You may want to have students make comparisons between the tidal ranges made near the head of the harbor or bay and those near the mouth. Observations made on consecutive days will reveal possible differences. Is there a funneling effect?

2. You may want to have students consult an *Ephemerus of Astronomical Observations* for the times of lunar apogee and perigee. Have students make their measurements on these dates and see how distance from the earth to the moon may affect the tidal range.

3. Have several students research how the Canadians are using the tides which occur in the Bay of Fundy for electrical power. Have the students report their findings to the class.

LAB REPORT

Long-Range Investigation 3
Tides at the Shoreline

Purpose
In this investigation, you will observe how the phases of the moon affect the tidal range.

Advance Preparation

1. Calculate the average distance to the water's surface for the following sample trials.

   a. High tide: Trial 1 2.5 m   b. Low tide: Trial 1 4.1 m

                 Trial 2 2.0 m   Trial 2 4.2 m

                 Trial 3 2.2 m   Trial 3 4.3 m

                         6.7/3 = 2.23 m   12.6/3 = 4.2 m

2. Calculate the tidal range of the average high- and low-tide measurements. 6.43 m/2 = 3.2 m

Observations

Table 3.1 Tidal Range

Date	Time	Lunar Phase	High Tide — Trial 1	High Tide — Trial 2	High Tide — Trial 3	High Tide — Average	Low Tide — Trial 1	Low Tide — Trial 2	Low Tide — Trial 3	Low Tide — Average	Tidal Range

Graph 3.1 Tidal Range

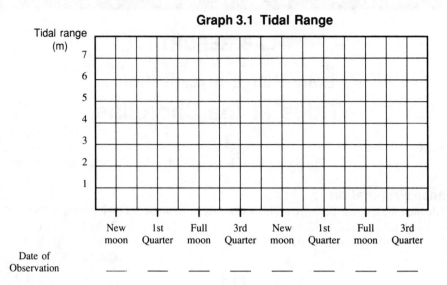

Tidal range (m)

7
6
5
4
3
2
1

New moon   1st Quarter   Full moon   3rd Quarter   New moon   1st Quarter   Full moon   3rd Quarter

Date of Observation ____ ____ ____ ____ ____ ____ ____ ____

Analysis and Conclusions

1. What causes the tides?

The gravitational pull of the sun and the moon.

2. **a.** During which phase(s) of the moon is the tidal range the largest?

The tidal range is largest (spring tide) during full moon and new moon phases.

   b. What was the tidal range?

Answers will vary.

3. **a.** During which phase(s) of the moon is the tidal range the smallest?

The tidal range is smallest (neap tide) during the first and last (third) quarter phases.

   b. What was the tidal range?

Answers will vary.

4. What local factors, if any, might affect the tidal range?

Answers may vary, but could include wind or storms, shape of basin and size of basin.

5. Describe the pattern of the graph. Did the pattern change in some regular, predictable way? If so, explain why this pattern exists.

The tidal range passes through a complete cycle of maximum to minimum and back to maximum in about 14 days. The pattern is predictable. The movement of the moon around the earth is constant. Because the moon affects the tides, the tidal range will have a pattern that is predictable.

Long-Range Investigation 4
Precipitation and the Water Table

Purpose
To discover how the level of the water table varies with the amount of precipitation.

Skills
analyzing, comparing, inferring, interpreting data, measuring, observing

Student-Textbook Correlation
Description of the water table is in textbook Chapter 14, Section 14.1 Water Beneath the Surface, pages 261–265.

Planning
Students will need two months to complete this investigation. Have students make their observations in early spring. If you wish to have students share their data and conclusions with the class, have them complete this investigation before you begin Chapter 14. The best months for this investigation will vary with the local climate. Try to schedule students when the local climate usually has a rainy period.

Cooperative Learning
1. Several students can be assigned different dates for more-continuous observations.
2. A variety of locations may be selected with different teams assigned to each location.

TEACHING STRATEGIES
Advance Preparation
Show students how to use or construct an accurate rain gauge.

Safety Guidelines
In this investigation, the following safety guidelines sould be emphasized.

CAUTION: Students should not go to the testing site alone. They should always go with their partner or an adult. They should wear a life jacket while working near water.

Possible Problems
Remind students to graph the distances to the water level downward so that the line graph will resemble changes in the water table.

Teaching Suggestions
1. Students may have difficulty in measuring the height of the pond water accurately. Suggest that they mark their reference point so that their measurements will be more precise.
2. Remind students to make scheduled visits to their testing site, as well as to record *all* precipitation. If you prefer not to have students observe the amounts of rainfall directly, have them copy the data from daily newspaper, television, or radio reports.

3. This long-range investigation may be completed as part of a field trip to a lake or pond. You may also want to tie in a discussion of lake (or pond) formation during the field trip as covered in Chapter 13, EarthBeat, page 245. A discussion of ecology, covered in Chapter 1, Section 1.1 What Is Earth Science, pages 7–8 can also be included.
4. Explain to students that groundwater usually flows down the slope of the water table, but because of pressure differences and differences in permeability of rock layers, groundwater flow patterns can vary.

For Further Investigation
1. You may want to have students compare the volume of water in a stream with rainfall in the area.
2. You may want to have students make a survey of the depths of local wells and include times when the water table is low and when it is high.
3. Have several students construct a monthly water budget showing precipitation and evapotranspiration for your region. Examine the monthly variations as they affect the water table.
4. Have several students research the "hardness" of the water in your area. Have the students explain to the class what effects water hardness has on water pipes and cooking utensils in the home.

M O D E R N   E A R T H   S C I E N C E

LAB REPORT

Long-Range Investigation 4
Precipitation and the Water Table

Purpose
In this investigation, you will discover how the level of the water table varies with the amount of precipitation.

Advance Preparation

1. What type of soil and rock is found in your area?

   Answers will vary.

2. What type of soil is best to hold water?

   Gravel and sandy soils.

3. What kinds of rock form good aquifers?

   Sandstone and limestone.

Observations

Table 4.1 Distance to Water Surface

Date	Distance to water level	Date of last rainfall	Amount of rainfall

Analysis and Conclusions

1. On which date was the water table the highest? the lowest?

Answers will vary.

2. Compare the date of the highest water table with the amount of rainfall during the preceding days or weeks. What does your comparison indicate?

Data will vary but should indicate that it rained a few days before the highest water table.

3. Compare the date of the lowest water table with the amount of rainfall during the preceding days or weeks. What does your comparison indicate?

Data will vary but should indicate that it was many days since the last significant rainfall.

4. Why might it take some time for rainfall to affect the water level in a lake or a pond?

Typically, ground water takes days or weeks to affect the water table.

Because water flows slowly underground, the water table will rise days after a heavy rain and drop slowly during a dry spell.

5. On hot days, some rainfall may evaporate before it soaks into the ground. Has the local temperature affected the water table? Explain.

Evaporation reduces the amount of rainwater that can raise the water table.

6. People generally use more water in summer than in winter. How might this affect the water table in an area where wells are the source of water?

Drawing out large amounts of water will lower the water table in the ground around a well, possibly causing wells to run dry temporarily.

7. If nearby hills were snow-covered when you began your observations, how would the melting snow have affected your results?

Melting snow releases water that has accumulated on the surface in cold weather. Some of the meltwater will seep into the water table. Spring thaws often raise the water table to its highest level of the year.

Long-Range Investigation 5
Air Pollution Watch

Purpose
To examine how the number of particles in the air may vary according to wind direction.

Skills
analyzing, comparing, drawing inferences, making and interpreting graphs, measuring, observing, understanding cause-and-effect relationships

Planning
Students will need two weeks to complete this investigation. Have students make their observation in early fall or late spring. If you wish students to share their data and conclusions with the class, have them complete this investigation before you reach Chapter 23.

Student-Textbook Correlation
Descriptions of major types of air pollution are found in Chapter 23, Section 23.1 Characteristics of the Atmosphere, pages 455–462.

Cooperative Learning
Different students could establish collecting sites at scattered locations or at varied altitudes. Such data might show that pollution is greater in different areas. Hypotheses could be established and tested.

TEACHING STRATEGIES
Advance Preparation
1. Compile a list of collecting sites for students that are in open areas where the wind can blow past the site from every direction. Check each site to be sure they are safe for the students.
2. Make arrangements with the biology teacher to use microscopes for counting particulates.
3. You may want to review with students how to make a bar graph.
4. You may want to have students completing this long-range investigation work with microscopes before they begin this investigation so that they can establish good microscope techniques.

Safety Guidelines
In this investigation, the following safety guidelines should be emphasized.

CAUTION: Injuries can result from improper techniques for driving a fencepost into the ground. Inexperienced students should be instructed to locate a post or platform that has already been set up. If they must set up their own post, they should request assistance from an adult.

Possible Problems
1. Students may need help in making accurate judgements of wind directions. You may wish to demonstrate this a few times before they begin the investigation
2. Students may need help in using the microscope and deciding which things observed on the slide are actually particulates and which are possibly blemishes in the grease film. They should examine the slides carefully.

Teaching Suggestions

1. Class time will need to be set aside for students involved in this investigation because most students will have access to a microscope only at school.
2. **a.** If your school is in a suburban area, you may want to have students set up their observation posts in their back yards.

   b. If your school is in an urban area, you may wish to have students set up their observations on a roof top, fire escape, or balcony. **CAUTION: Emphasize to students that roof tops and fire escapes can be dangerous. Remind them not to set up their investigation near the edge of the roof or fire escape.**
3. Have students place several slides on a windowsill before actually beginning the investigation. Students can make counts from these slides to become familiar with the techniques of greasing and counting the particulates on the slides.
4. If a certain wind direction seems to correlate with the highest particulate counts, investigation stations might be set up at varying intervals along this direction. The purpose would be to locate the source of the pollutants.

For Further Investigation

1. Set up this same investigation during different seasons of the year. Correlate amounts of particulates with weather events and seasonal climate variations. Do this for various weather events as well.
2. You may want students to interview community leaders for suggestions about causes of local air pollution. Find out if there has been an increase or decrease in local air pollution in recent years, and if so, what factors contributed to the changes.

M O D E R N   E A R T H   S C I E N C E

LAB REPORT

Long-Range Investigation 5
Air Pollution Watch

Purpose
In this investigation, you will examine how the number of solid particles in the air may vary according to wind direction.

Observations

Table 5.1

Week of _____

Day	Date	Wind direction	Slide direction	Number of particulates 1	2	3	4	5	Total
1			N						
			S						
			E						
			W						
2			N						
			S						
			E						
			W						
3			N						
			S						
			E						
			W						
4			N						
			S						
			E						
			W						
5			N						
			S						
			E						
			W						
6			N						
			S						
			E						
			W						
7			N						
			S						
			E						
			W						

HRW material copyrighted under notice appearing earlier in this work.

133

Graph 5.1

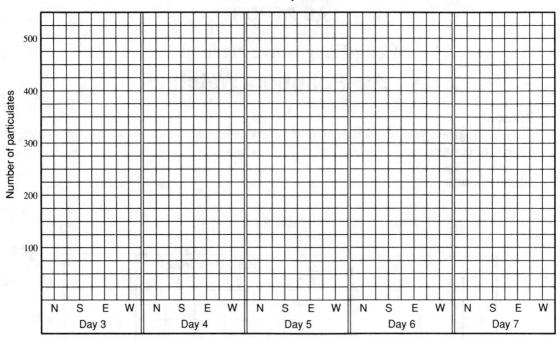

Analysis and Conclusions

1. For your location, is there any one wind direction or group of wind directions that resulted in more particulates than any others?

   Answers will vary depending upon local conditions.

2. Could you identify any of the particulates, such as soot, pollen, pieces of leaves, ash, and so on? If so, list some of the identifiable particulates.

   Answers will vary.

3. What are possible sources of these particulates?

   Answers will vary depending upon local conditions such as proximity

   to industry, highways, and so forth.

4. What would your results likely be if you set up this investigation near a populated urban area?

   Numbers of particulates would most likely be higher due to heavy traffic

   and soot from buildings. Wind direction might also be affected by tall

   buildings. Answers will vary.

5. Did weather conditions have any effect on your data results or patterns? Explain.

   Answers will vary.

Long-Range Investigation 6
Correlating Weather Variables

Purpose
To examine how the relationship among certain weather variables helps to predict the weather.

Skills
analyzing, comparing, inferring, interpreting graphs, measuring, making generalizations, predicting

Planning
Try to plan this activity so that it will culminate when you are about to start your weather unit. If you plan to have students share their work, data, and conclusions with the class, be sure the investigation is completed before you begin Chapter 25. Students will need approximately one month to complete this investigation.

Student-Textbook Correlation
Descriptions of water in the atmosphere are found in Chapter 24, pages 478–493.
Descriptions of other atmospheric variables are found in Chapter 25, Sections 25.2 Fronts and 25.3 Weather Instruments, pages 502–511.

Cooperative Learning
Teams of students can share the responsibilities for measuring different variables at the same time or the same variables at different times.

TEACHING STRATEGIES
Advance Preparation
1. If possible, secure the weather instruments ahead of time and set up a formal weather station. However, this investigation can effectively be completed without the use of sophisticated instruments, which may not be available.
2. Have students observe clouds a few days before they begin this investigation. Have them determine if the types of clouds they observe on a given day are associated with certain weather conditions. Ask them if clouds give clues to the weather conditions of the next day.

Possible Problems
1. Some weather instruments are fragile and easily damaged.
2. Frequent reminders to collect data may be necessary.
3. Interpretations of amount of cloud cover may vary.
4. Many science textbooks use only metric measurements. It must be noted that the National Weather Service reports temperatures in degrees Fahrenheit, precipitation amounts in inches, and wind speed in miles per hour or knots. Barometric pressure is plotted on maps using millibars while it is reported in the news media in inches of mercury. Pressure in the investigation is indicated using millibars. Students will be able to convert inches to millibars using Figure 6.1 if necessary.

Teaching Suggestions

1. Explain to students that the purpose of this investigation is to determine if weather patterns exist and what those patterns are.
2. Remind students to be sure that the temperature is measured in the shade and in an area that is protected from the wind.
3. The causes of the relationship between pairs of weather data should be explained as a follow-up to the investigation.
4. A large replica of Graph 6.1 placed on a bulletin board will prove useful for class discussion.
5. Make a concentrated solution of cobalt chloride and water. With a brush, paint the solution onto one surface of a blank piece of paper. When the paper dries it can be used to indicate the amount of moisture in the air. It will turn blue when the humidity is below 50% and pink when the humidity rises above 50%.

For Further Investigation

1. You may want to have students identify the causes for the changes in weather variables preceding rain.
2. Have students research the conditions associated with low-pressure areas that are necessary for hurricane formation.
3. You may want to have students investigate what weather patterns associated with the Plains States lead to the formation of tornados in the late spring.

M O D E R N E A R T H S C I E N C E

LAB REPORT

Long-Range Investigation 6
Correlating Weather Variables

Purpose

In this investigation, you will examine how the relationships between certain weather variables aid in the prediction of the weather.

Observations

Table 6.1

Day	Temperature (°C)		Barometric pressure (mb)		Wind direction		Cloud cover		Present weather	
	A.M.	P.M.	A.M.	P.M.	A.M.	P.M.	A.M.	P.M.	A.M.	P.M.

Analysis and Conclusions

1. Referring to your graph, on how many days was the temperature falling?
   Answers will vary.

2. How many of the days with falling temperature also had rising barometric pressure?
   Answers will vary. However, there is usually better than a 50% chance.

3. In general, what is the relationship between temperature and pressure?
   Usually these variables tend to be inversely related, but the exceptions are numerous.

4. How many days had falling barometric pressure?
   Answers will vary.

5. What sky cover is usually associated with falling pressure?
   Answers will vary, but more than 50 percent of cloud cover is typical.

6. What wind direction is usually associated with falling barometric pressure?
   Answers will vary, but easterly winds usually accompany falling pressures.

7. What weather conditions are usually associated with high barometric pressure?
   A typical response is cool temperatures, clear skies, and westerly winds.

8. What weather conditions are usually associated with low barometric pressure?
   A typical response is warm temperatures, cloudy skies, and easterly winds. Also, storms generally occur when the barometric pressure is low.

9. How do the relationships between certain weather variables help to predict the weather?
   If the wind direction changes to easterly or the pressure is falling, stormy weather is likely to come soon. Conversely, fair weather can be predicted when the winds shift toward the west and the barometric pressure rises.

Long-Range Investigation 7
Weather Forecasting

Purpose
To use a series of daily weather maps to track the movements of weather systems and then make a prediction of future weather conditions for a given location.

Skills
analyzing, comparing, inferring, interpreting data, predicting

Planning
Students will need two weeks (one in winter and one in spring) to complete this investigation. If you plan to have students share their data and conclusions with the class, this investigation must be completed before you begin Chapter 25.

Student-Textbook Correlation
Descriptions of weather maps are found in Chapter 25, Section 25.4 Weather Forecasting, pages 512–517. Additional information on weather can be found in Chapter 23, Section 23.3 Winds, pages 469–473 and Chapter 25, Section 25.1 Air Masses, pages 499–501.

Cooperative Learning
Groups of students can be assigned different weeks so that a continuous weather record is maintained.

TEACHING STRATEGIES
Advance Preparation
1. Make a list of local newspapers that contain a useful daily weather map. National newspapers contain weather maps that can also be used by students.
2. There are several other sources for daily weather maps.
   a. Local airports have weather-service ''facsimile'' maps, which are often available to teachers after they have been used.
   b. Large daily maps can also be obtained from:
   > Superintendent of Documents
   > Government Printing Office
   > Washington, DC 20402

   The minimum subscription period is three months and a nominal fee is charged.

Possible Problems
1. Students may have difficulty in locating pressure centers on the weather map. Encourage students to use state borders as guides in locating these centers.
2. Some students may encounter problems with the computations of daily rates of movement of pressure centers.

3. Students will be concerned about being ''wrong'' with their forecasts. Emphasize that there is always a certain percentage of error involved with forecasting weather, especially because interpreting daily maps uses averages. Comparison with Long-Range Investigation 6, "Correlating Weather Variables," may be helpful.

Teaching Suggestions

1. In order to make a long-term weather prediction, it is necessary to use student information on the rate of motion and the track of their pressure system. By daily observation of the student's maps, you can help the students determine both the speed and direction of one or more weather systems as they move across the United States.

2. Explain that a low-pressure system usually moves across North America with the prevailing westerly winds. The direction of movement of the low-pressure system tends to be across isotherms and parallel to isobars. Strong winds will tend to slow the motion of a low-pressure center. Emphasize that often there can be significant variations from the average velocity when these strong winds exist. This information may help students explain why the movement of the pressure systems they are tracking are not always consistent.

3. Explain that high pressure systems tend to move toward the area where the greatest increase in barometric pressure is occurring. A well-developed high-pressure system, which is east of a low, will tend to slow the low's movement or deflect it to the left or right. Also, two low-pressure centers tend to merge.

For Further Investigation

1. You may want students to list the factors responsible for the direction in which low- and high-pressure centers move across North America from day to day.

2. Have students investigate how a wave cyclone, with its associated fronts, develops in North America.

3. Have interested students research how the jet stream influences the weather patterns in the United States and Canada.

M O D E R N   E A R T H   S C I E N C E

LAB REPORT

Long-Range Investigation 7
Weather Forecasting

Purpose
In this investigation, you will use a series of daily weather maps to track the movements of weather systems and then make a prediction of future weather conditions.

Advance Preparation
1. You will notice a variety of symbols on your weather maps. What do these symbols represent?

   These symbols represent wind direction and speed, weather fronts, and

   pressure areas.

2. Are the symbols used on the different weather maps similar? Why?

   Yes. So that meteorologists all over the world will understand any weather

   map that they observe.

3. **a.** How is a low-pressure system indicated on the weather map in the newspaper?

   By using a capital L.

   b. How is a high-pressure system indicated on the weather map in the newspaper?

   By using a capital H.

Observations

Table 7.1 Local Weather Predictions

Your hometown weather predictions										
	Week One (winter)					Week Two (spring)				
	1	2	3	4	5	1	2	3	4	5
Temperature										
Barometric pressure										
Barometric trend (R = rising F = falling S = steady)										
Wind direction										
Cloud cover (C = clear cl = cloudy PC = partly cloudy O = overcast)										
Present weather (rain, sleet, snow, etc.)										
Prediction										

Analysis and Conclusions

1. In what general direction do the pressure centers over the United States move?

The general direction is west to east.

2. From your calculations, what is the average rate of movement, in kilometers per day, of low- and high-pressure centers in winter?

Answers will depend on students' maps.

3. Predict where the low- and high-pressure centers will be located on the day following the date of the last map in your series.

Locations will depend upon the map sequence for the period of the

investigation.

4. Describe the general weather conditions associated with regions of low and high atmospheric pressure.

Low-pressure regions have a relatively warm temperature, increased cloud

cover, high humidity, surface winds flowing in a counterclockwise pattern,

and high probability of precipitation. High-pressure regions have a relatively

cool temperature, clear skies, low humidity, surface winds flowing in a

clockwise pattern, and low probability of precipitation.

5. Based on your series of daily weather maps, predict the weather for your hometown on the fifth day of the series. Write a forecast and fill in the data table with your estimates of weather conditions.

Predictions will depend upon weather patterns during the course

of the investigation.

6. Compare your prediction with the daily weather map for the fifth day. Check the accuracy of your prediction. What factors could have resulted in errors in your prediction?

Factors that can result in errors in predictions are changes in the rate of

movement of the highs and lows from the calculated average, a shift in the

path of the pressure systems, and a shift in the position of the jet stream.

7. From your calculations, what is the average rate of movement, in kilometers per day, of low- and high-pressure centers in the spring?

Answers will vary but average rates should generally be slower in winter.

8. After completing both winter and spring observations, compare the rate of movement of pressure systems during the different seasons.

Generally, pressure systems have a greater rate of movement in the winter

than in the late spring.

9. How can you predict weather using a series of daily weather maps?

By observing the direction and rate of travel for pressure centers, you can

calculate the approximate arrival time of weather coming from the west.

Long-Range Investigation 8
Comparing Climate Features

Purpose
To have students use climate data to compare their local climate with other regions of the United States.

Skills
analyzing, classifying, comparing, drawing inferences, making and interpreting graphs, using special references

Planning
This investigation will take six months to complete, from October through March. When you reach Chapter 26, you might have students relate what their observations have indicated thus far. Different groups of students could be assigned different days to monitor the local and national weather conditions. The groups could then share the data so that daily observations by each group would not be necessary.

Student-Textbook Correlation
Types of climate are described in Chapter 26, Section 26.2 Climate Zones, pages 529–535.

Cooperative Learning
1. If students choose to measure temperature and amount of precipitation, data could be collected by several students and averaged. Not only would this tend to produce more reliable observations, but it would also account for any missing items of data.
2. A group of students could prepare climatographs for a large number of cities. Each member of the group should prepare climatographs for a few cities and share this information with other members of the group or the class.

TEACHING STRATEGIES
Advance Preparation
1. Each student will need access to an almanac and an atlas. These can be made available in the classroom or obtained from the library.
2. If applicable, check to see if the students know how to use and to read thermometers and rain gauges. Have several students make the same observation by using a thermometer and rain gauge, and look for differences in the readings.

Possible Problems
1. Students may be confused in estimating amount of daily precipitation when rain continues for more than one day. Suggest that they divide the entire amount of precipitation into portions that are proportional to the hours of rainfall each day.
2. Because such a large file of daily data is collected, it is helpful to check student logs at the end of each month. The calculation of monthly data provides an opportunity for feedback. Periodic reminders are necessary to keep students from forgetting to make observations.

Teaching Suggestions

1. If direct measurements will be taken, secure a location on the school grounds for students to observe temperature and precipitation. The instruments should be kept safe from theft or vandalism. An ideal arrangement would be to use a permanent weather station.
2. Maximum–minimum thermometers are best for recording temperatures in this investigation. If maximum–minimum thermometers are available, students should be instructed in the proper use of the thermometer.
3. Arrangements should be made to permit students to record data at appropriate times during the school day.
4. On days when direct measurements cannot be made, students should record data taken from newspaper, television, or radio reports.
5. A classroom calculator or computer may ease the task of monthly averaging.

For Further Investigation

1. Have students investigate why the coast of northern Chile is a desert in the easterly trade-wind belt and the coast of southern Chile, in the belt of the prevailing westerlies, is quite humid. Have students compare the Chilean coastal climate with the climate of the west coast of the United States.
2. You may want to have students research why, around the world, people tend to cluster in certain regions, while others are sparsely populated. Which climate systems are associated with dense populations? Give some possible reasons for such distributions.
3. You may want to have students discuss why, in the United States, the centers of population are shifting toward the south and west. Suggest some reasons for this trend. List some problems associated with the change of population density.

M O D E R N   E A R T H   S C I E N C E

LAB REPORT

Long-Range Investigation 8
Comparing Climate Features

Purpose
In this investigation, you will use climate data to compare your local climate with that of other regions of the United States.

Observations

Regional climate chart for _____

	J	F	m	a	M	j	J	A	S	O	N	D
Temperature (°)												
Precipitation ( )												

Climatograph

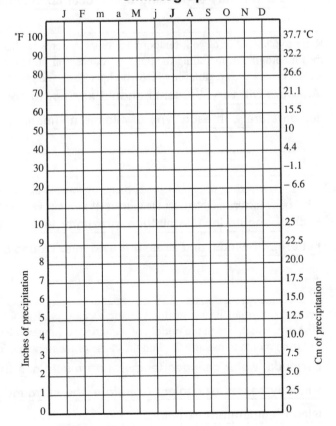

Key to months:
a	Apr	M	May
A	Aug	F	Feb
J	Jan	S	Sept
j	Jun	D	Dec
J	Jul	N	Nov
m	Mar	O	Oct

Figure 8.1

Analysis and Conclusions

1. Compare each of the seven climatographs that you prepared with the sample climatographs for the major world climates on page 52. Identify the climate type or combination of climates for each location you selected. Explain the features of your climatograph that helped you classify each region.

   The New England region has a humid continental climate. The Gulf Coast

   region has a subtropical climate. The Midwest region has either a steppe or

   humid continental climate depending on the city selected. The Southwest

   has either a desert or steppe climate, depending on proximity to the west

   coast mountains. The Pacific west coast has either a Mediterranean or

   marine west coast climate. Interior Alaska has a subarctic climate. The

   Hawaiian Islands have both tropical rain forest and savanna climates.

2. Using the climatograph for your locality, classify your regional climate. Explain the features of your climatograph that helped you choose the climate type. From your experience in previous years or from information found in reference sources, explain whether you think this year's data is typical for your region.

   Answers will vary depending upon local conditions.

3. Refer to Chapter 26 of your textbook for other systems to classify climates. How would each of the climatographs, including the one for your region, fit into such systems? Using the new systems, classify each region for which you have data, and give your reasons.

   Answers will vary. Students should attempt to correlate rainfall and

   temperature with each type of climate mentioned.

4. In this investigation, you compared climates by looking at average precipitation and temperatures. What other factors might affect the climate of an area? Give examples to illustrate each factor.

   Factors should include altitude, proximity to large bodies of water,

   patterns of winds, and pressure systems. Examples will vary.

5. Bermuda, a small island in the Atlantic Ocean, is at about the same latitude as St. Louis, Missouri, which lies in the middle of a continent. In which location does the temperature vary less from month to month? Explain the cause of the more moderate temperature pattern.

   Bermuda; water changes temperature much less than land that has been

   similarly heated or cooled. Therefore, the surrounding water moderates

   island temperature variations.

Long-Range Investigation 9
Planetary Motions

Purpose
To observe the motion of the planet Mars against the background stars over a period of several months. Conclusions about the orbit of Mars can then be formulated.

Skills
analyzing, evaluating, inferring, interpreting data, observing, predicting

Planning
If you plan to have students share their work, data, and conclusions with the class, be sure the investigation is completed before you begin Chapter 29. Students will need nine months to complete this investigation, therefore this investigation should be assigned at the beginning of the school year.

Student-Textbook Correlation
Descriptions of the planet Mars are provided in the textbook, Chapter 29, Section 29.2 The Inner Planets, pages 594–597.

Cooperative Learning
Students could be assigned staggered schedules to produce a more detailed graph.

TEACHING STRATEGIES
Advance Preparation
1. Make copies of the constellation chart to be given to the students.
2. Explain to students how to locate constellations before this investigation starts. It is best for you to become familiar with the constellations and where to find them in the night sky before you attempt to explain this to your students.
3. Prepare a list of periodicals and newspapers which give accurate information on the time and location Mars will be visible in the sky. Distribute the list to the students.
4. Supply students with information on Mars's position in the sky for last year so they can compare that position with the present position. This information can be obtained from a planetarium, college astronomy department, or a local astronomer.

Safety Guidelines
In this investigation, the following safety guideline should be emphasized. Students should be cautioned to dress warmly for making observations on winter or cool nights. Students should also have a partner or an adult with them.

Possible Problems
1. Poor weather may obscure the sky.
2. Students may forget to make their observations on the correct days of the month. They should be reminded to keep to the schedule.
3. Students may have difficulty finding Mars in the nighttime sky, and then locating it on their constellation charts.

Teaching Suggestions

1. Explain that Mars has a longer period of revolution around the sun than does the earth. Periodically the earth overtakes Mars and passes it in an easterly direction. As a result, for a period of months Mars appears to move backwards in a westerly direction against the stellar background. This is referred to as *retrograde* motion.
2. Explain to your students that the synodic period of a planet's revolution is the time required for the planet to return to the same position in the sky as seen from the earth. The synodic period of Mars is two years forty-nine days.
3. Organize a field trip to a planetarium so students can observe the constellations and be able to find them in the nighttime sky.
4. You may wish to have several students make observations of Jupiter or Saturn. At the completion of the investigation, the students should share their observations with the class and make comparisons with Mars.

For Further Investigation

1. You may want students to research if Mars appears to have retrograde motion each year. If not, Why not?
2. Have students investigate the synodic period of Mars.
3. You may want to have students discuss the general direction of motion across the sky of all the planets.
4. Have interested students compare Mars with the earth pointing out the differences and similarities.
5. You may wish to have students investigate why the planet Mars was named after the Roman god of war.

M O D E R N   E A R T H   S C I E N C E

LAB REPORT

Long-Range Investigation 9
Planetary Motions

Purpose
In this investigation, you will observe the planet Mars over a period of several months and use your observations to draw conclusions about planetary motion.

Analysis and Conclusions

1. In which months does Mars appear highest in the night sky? lowest?

   Mars appears highest in September and April and lowest in March and June.

2. In which direction does Mars appear to move across the sky?

   Mars appears to move from east to west, but for a few months it

   appears to be moving backwards.

3. Describe the apparent path of motion of Mars plotted on your star chart.

   Mars appears to move in an erratic line across the star chart.

4. Will Mars be in the same position one year from today? Explain.

   No, since it takes Mars two years to orbit the sun, the differences in the

   orbits will not make Mars appear to be in the same position one year from today.

5. Referring to your observations of the apparent brightness of Mars throughout the investigation, what can you infer about the distance between Mars and the earth?

   The distance between Mars and the earth varies because the brightness

   of Mars changes during the year.

Figure 9.2 Constellation Chart
(Declination −50° to +50°)

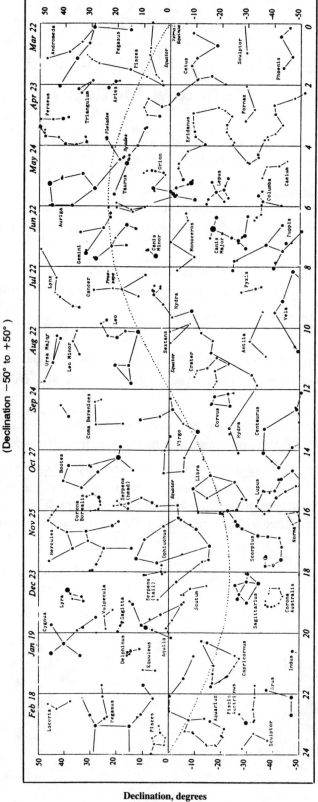

Long-Range Investigation 10
Apparent Motions of the Moon

Purpose
To observe how the moon's appearance and altitude vary from one new moon to the next full moon.

Skills
analyzing, comparing, constructing, inferring, making and interpreting graphs, measuring.

Planning
Students will need one month to complete this investigation. If you plan to have students share data and conclusions with the class this investigation should be completed before you begin Chapter 30. It would be best to have this investigation begun in the fall or early spring when the temperature is moderate.

Student-Textbook Correlation
Description of the moon's phases and motions is found in the text Chapter 30, Section 30.3 The Lunar Cycle, pages 626–629.

Cooperative Learning
Deviations in altitude and azimuth measurements can be reduced by averaging data recorded by different observers.

TEACHING STRATEGIES
Advance Preparation
1. Before assigning this investigation explain how altitude and azimuth coordinates are used to locate planets and other celestial objects.
2. Students may need assistance in assembling and using the sextants.
3. Review of the use of a magnetic compass is advised.

Safety Guidelines
In this investigation, the following safety guideline should be emphasized. Students should be cautioned to dress warmly for making observations on winter or cool nights. Students should also have a partner or an adult with them.

Possible Problems
1. There may be cloud cover obscuring the moon for several days. Be sure to tell students that if a cloud cover is evident on their observation day that they should indicate it in their data table.
2. Students often forget to make their observations. Reminders and a printed schedule of observation dates is advised.
3. Students may accidentally move the string of their "sextant" causing an incorrect angle during their observations.

HRW material copyrighted under notice appearing earlier in this work.

151

Teaching Suggestions

1. If poor weather conditions prevent observations for a significant portion of the month, the activity should be repeated the next month.

2. Observations every night are ideal. However, this may not be practical. During the first two weeks, a schedule of observations on at least the following nights at 7:00 P.M. (or just after sunset) is recommended: night 3 or 4 (waxing crescent), night 7 or 8 (first quarter), night 14, 15, or 16 (full moon). Observe the moon later in the evening for nights 18 to 21 at about 9:00 P.M. (waning gibbous). Early morning observations (about 7:00 A.M.) are recommended for days 25 to 27 (waning crescent). Note that these days are after the new moon and are approximations. You can guide students' observations by referring to a local paper or a moon-phase calendar. If the Internet is available, you can check the current week's phases at stardate.utexas.edu/nightsky/astroupdate.html#moon.

3. If a telescope is available, encourage students to view the moon and its details. The most satisfactory time to view the moon is in one of the partial phases because the full moon is too bright for detailed viewing.

4. Explain to students that the moon rises later on successive nights because of its position on its orbit. Ask students to find the average delay of moonrise on successive days.

For Further Investigation

1. Have students explain why, during the crescent phases of the moon, the dark side of the moon is faintly visible.

2. Ask a group of students to research why the period from new moon to the next new moon is two days longer than the moon's period of revolution around the earth.

3. You may want students to explain why the same side of the moon always faces the earth.

4. Have students hypothesize how erosion occurs on the moon.

M O D E R N   E A R T H   S C I E N C E

LAB REPORT

Long-Range Investigation 10
Apparent Motions of the Moon

Purpose
In this investigation, you will observe how the moon's appearance and altitude time vary from one new moon to the next new moon.

Observations

Table 10.1 Time of Observation _____

Date	Phase	Altitude (°)	Azimuth (°)

HRW material copyrighted under notice appearing earlier in this work.

153

Graph 10.1 Position of the Moon in the Sky

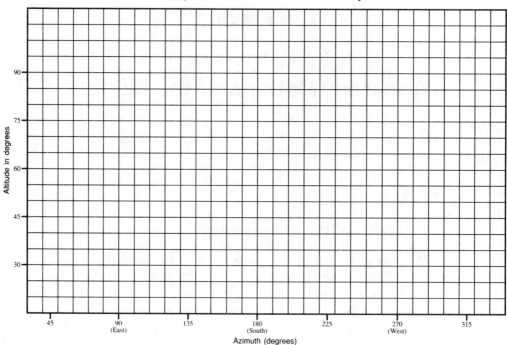

Analysis and Conclusions

1. What sequence of the moon's phases did you observe?

<u>Answers should include some part of this sequence: new moon, waxing</u>

<u>crescent, first quarter, waxing gibbous, full moon, waning gibbous, last</u>

<u>quarter, and waning crescent phases.</u>

_____

2. a. You observed the moon each night at about the same time. Referring to your graph, what did you notice about the position of the moon during each observation?

<u>The moon is observed farther east each night.</u>

b. What might you conclude about the time of moonrise from night to night?

<u>Moonrise occurs later each night during this sequence of phases.</u>

3. Describe the changes in the moon's appearance from the new moon to the full moon.

<u>The entire surface facing the earth is dark at new moon. As the phases</u>

<u>change, beginning with the right (east) edge of the moon, a greater portion</u>

<u>of the moon's surface is in sunlight. At full moon, the entire surface facing</u>

<u>the earth is lighted.</u>

4. How does the moon's appearance and altitude at a particular time vary from one new moon to the next?

<u>The appearance is the same. However, the altitude changes from one month</u>

<u>to the next.</u>

About the Materials and Equipment List

The table on the following pages lists the materials and equipment needed for the entire lab and field program for *Modern Earth Science*. Included are materials for Small-Scale Investigations, In-Depth Investigations, and Long-Range Investigations. The investigation number or chapter number is provided in separate columns for easy reference. Suggested quantities per student or lab group are given in the last column. The amount listed is the maximum quantity needed to complete each investigation. Where no quantity is listed, it is assumed that materials are common and readily available.

Before ordering any of the materials and equipment listed here, you may want to review the individual investigations you plan to assign. By doing this, you will become familiar with the procedures and will be able to order materials based on your class needs. For example, you may want to separate your class into groups rather than have individuals perform each investigation. Also, a review of the materials in context will prove helpful when ordering bulk materials such as clay, soil, and gravel.

The materials and equipment used in the various investigations may be purchased from WARD'S or another scientific supply house. With WARD'S exclusive computerized ordering system for *Modern Earth Science,* you can order your materials and supplies quickly and easily. Many of the materials found in the list, however, can be easily made at home or in school, or they may be purchased in local hardware and office-supply stores. If students are using apparatus from the school to perform the Long-Range Investigations, they should sign out for the materials or equipment so that you will know who has the materials or equipment and when to expect their return.

HRW material copyrighted under notice appearing earlier in this work.

155

Materials	WARD'S Catalog No.	Small-Scale	In-Depth	Long-Range	Quantities per student or group
adding-machine paper	27 M 3076	18	5, 18		5 m; 35 cm, 5 m
almanac				8	1 per class

aluminum foil
   12 cm × 12 cm
   4 cm × 4 cm
	15 M 1009	
		
		28
			
			27
			28
					
					1 piece
					1 piece

ammonia solution	37 M 0416		12		1–2 mL
aneroid barometer	23 M 1604			6	1
atlas				8	1 per class
balance, metric	15 M 6057	6	8, 20		1

balloon
   round (large, 6–7 cm)
   small
	15 M 0017	
		1
		8
					
					1
					1

basin, flat (large, at least 8 cm deep with a flat bottom)	20 M 3210		3		1

batteries
   AA
   6 V dry cell
	
	15 M 3256
	15 M 3263
			
			27
			8
					
					3
					1

beads, plastic
   4 mm
   8 mm
	
	36 M 4195
	36 M 4196
			
			14
			14
					
					400
					200

beakers
   100 mL
   250 mL
   400 mL
   1 L or larger
	
	17 M 4020
	17 M 4040
	17 M 4050
	17 M 4080
			
			14
			21
			8, 23
			4
					
					1
					3
					1
					1

bearing chart				1	1
bobby pins (long)			5		5
books		5, 8			2
box, cardboard (20 cm × 30 cm)		15, 16			1
brick		13			1
Bunsen burner with flame spreaders	15 M 0612 15 M 0607		11, 21 4		1 2
calculator	27 M 3055		6, 30		1

cardboard
   10 cm × 25 cm
   20 cm × 30 cm
   flexible (3 cm × 30 cm)
   shirt-packaging
   thick
		
		23
		
		2
		19
		
			7
			
			5
					
					1 piece
					1 sheet
					1 strip
					8 1-cm squares
					15 cm^2

celestial sphere	80 M 8510			9	1
cereal flakes		7			8 oz.

Materials	WARD'S Catalog No.	Small-Scale	In-Depth	Long-Range	Quantities per student or group
cheesecloth	15 M 0015	14 24			15 cm^2 5 cm^2
clock or watch (with second hand)	15 M 1492	10, 11, 13, 14, 16, 17, 25, 26	2		1
cloth cotton (5 cm $\times$ 5 cm) wool (15 cm $\times$ 15 cm)	15 M 2537	8	24		1 piece 1 piece
cloth ties (50 cm long)			22		2
coffee can (10-cm diam.)	17 M 2111	23			1
comb, rubber or plastic		8			1
compass drawing magnetic	15 M 4648 12 M 0601	29	6 2	1, 5, 6, 9, 10	1 1
constellation chart	33 M 1410			9	1
container, waterproof (large, at least 8 cm deep)	21 M 2100	3			1
copper (cupric) carbonate	37 M 2214		11		2 g
copper penny		9	9		1
cork stoppers	15 M 8364 15 M 8366	10, 22	12		9 3, 1
craft sticks	15 M 9893		4		2
distilled water	88 M 7005	21			1 L
dowel, soft-wood (2 mm $\times$ 15 cm)	15 M 0014	5			1
dowel, wooden 1/4 in. diam. (12 in.) thick	15 M 0081 15 M 0082		2 3		1 1
drinking glasses; jars 4 oz. 16 oz.	17 M 2090 17 M 2098	8 10			1 3
drinking straw, plastic	15 M 9869	21, 23	21		1
electrodes, stainless steel	16 M 0501		8		2
Epsom salts	37 M 2860	10	8		120 mL (1/2 cup); 300 g
eraser (rubber)					1
eyedropper	17 M 0230		7		1
fan cardboard (small) electric, 3-speed	15 M 9854	24 25			1 1

Materials	WARD'S Catalog No.	Small-Scale	In-Depth	Long-Range	Quantities per student or group
felt-tipped pen		6			1
water-based	15 M 1151	1			1
filter mask	15 M 3061	16			1 per student
flashlight	15 M 3264		26	9, 10	1
flashlight bulbs (3 V)	16 M 0541		27		2
food coloring (dark)	15 M 0071	19	4		1 bottle
freezer		15, 21	15, 21		1
funnel	18 M 1421		11		1
gelatin, red-colored			7		16 mL
glass plate/square	12 M 0008	9	9		1
gloves					
heat-resistant	15 M 1095		11, 21		1 pair
latex	15 M 1071		8		1 pair
glue	15 M 9806			1	1 bottle
glycerin	39 M 1438		23		1 bottle
graduated cylinder					
50 mL	18 M 1720	26	21		1
100 mL	18 M 1730	7	13, 14		1
gravel	45 M 1990	14			1 cup
	45 M 1985	15			1/4 cup
grease pencils	15 M 1155		13, 20 21	5	1 2 (1 yellow, 1 red)
hair dryer		16			1
hammer	12 M 0110			5	1
hand drill				1	1
hand lens	24 M 1112	12	1, 9, 10		1
hand towel		15			1
heat lamp	36 M 4168 36 M 4173		26		1
hot plate/stove	15 M 7999	10, 19			1, 2
hydrochloric acid, dilute	750 M 3915		10		2 mL
10%	37 M 9561		12		10 mL
ice		10	23		4–6 large cubes
index cards	15 M 9819	28	28		2; 1
iron filings	37 M 2312		11		5 g
jar(s), glass					
1 L	17 M 2098	10, 21			3, 1
clear	17 M 2051	12			1
with lid	17 M 2153		28		1

Materials	WARD'S Catalog No.	Small-Scale	In-Depth	Long-Range	Quantities per student or group
juice containers, cardboard (12 oz.)			13		2
knife, plastic	250 M 8128	4	3		1
lab apron	15 M 1005	4	4, 7, 8, 11, 12, 17, 29		1 per student
ladder		27			1
lamp or other light source desk (with incandescent bulb)	36 M 4168		27, 28		1
portable (with clamp or flexible neck and incandescent bulb)	36 M 4173	26			1
lead sinkers (100–200 g)	350 M 0237			2, 3	3, 1
leaf			17		1
marbles	15 M 3399	2	29		1; 6 (1 shooter)
markers	15 M 1151	16, 19, 21	22, 29		1
matches	15 M 9427		8		1 box
measuring cup	15 M 9873	7, 10, 14, 21			1
medicine dropper	17 M 0230		10, 12		1
meter stick	15 M 4065	18, 20, 26, 27	1, 18, 22, 26, 29	2, 3, 4	1
microscope	24 M 2310			5	1
microscope slides	14 M 3500			5	8
milk carton (small); (large)			15, 16		1, 2
mineral samples: 46 M 1192, 46 M 3822, 46 M 3862, 46 M 4847, 46 M 8002			9		5
mineral specimen: calcite (46 M 1422), feldspar (46 M 5122), galena (46 M 3332), graphite (46 M 3702), gypsum (46 M 3792), mica (46 M 1992), pyrite (46 M 6447), quartz (46 M 6547)		9			1
modeling clay	36 M 4147	3 4 15 21 30	3 17 21 28		1–2 lb.; 4 lb. 2–3 lb.; 1/4 lb. 1–2 lb.; 1–2 oz. 1–2 oz.; 2–4 oz. 2–4 oz.
nail (large)	15 M 9478		13		1
nylon cord	15 M 2543			2, 4	13m, 11m
pad of paper		6			1
paint black (flat finish)			28	2	2 pints (1 black, 1 white) 5–10 mL

Materials	WARD'S Catalog No.	Small-Scale	In-Depth	Long-Range	Quantities per student or group
pans					
23 cm × 33 cm	14 M 7010	13, 19, 25	4, 13		2,1,1; 1, 1
30 cm × 40 cm × 10 cm	15 M 0539	22	15		1
cooking (small)	14 M 6997	11			1
paper					
2 m × 1 m			22		
carbon			17		
construction	15 M 9825		26		
graphing	15 M 3835		22, 23, 26	1, 2, 3, 4, 5, 6, 8, 10	
newspaper		6	17		6–9 sheets
notebook		4, 17, 18, 26, 29, 30	19	1	
opaque			3, 17, 25		
white		2, 29			
paper bag (large)		16			1
paper clip	15 M 9815	3		3, 10	1
paper cups	15 M 9830		20		5
paper fastener (1 1/4 in.)				1	1
paper napkins			5		2 (1 dark, 1 light)
paper plates	15 M 9889				
large		7			2
small (15 cm in diam.)		27			5 (1 red, 4 blue)
paper towels	15 M 9844	13	20		1 roll
paraffin bricks	250 M 0817		27		2
(12 cm × 6 cm)					
pebbles (small)	45 M 1986	15	15, 16		1/2 lb.; 1/2 lb., 1 lb.
pencils	15 M 9816	3	2, 3, 4, 17, 18, 19, 25, 28		1
colored	15 M 4690	18	22	7	set of 6; 3; 2
			25		2 (1 red, 1 blue)
sharpened		14	7		1
penlight	15 M 4000	30			1
Petri dishes (small)	18 M 7101	26			3
petroleum jelly	15 M 9832		23	5	1 jar
pH paper	15 M 2558		12		6 pieces
plaster of Paris	37 M 2149	7	16, 17, 29		1/2 cup; 500 g
plastic, black		11			1 small sheet
plastic bag	18 M 6952	8			1
(large)	18 M 6995	6			1

Materials	WARD'S Catalog No.	Small-Scale	In-Depth	Long-Range	Quantities per student or group
plastic column, clear (32 in. × 1.5 in.)	36 M 4193		20		1
plastic containers					
15 cm × 10 cm × 5 cm	20 M 3250	15			1
16 oz. or larger	18 M 9917	12	17, 24		1
with lid	18 M 9917	12			1
bowl (8 oz.)	18 M 7203		26		2
tray (large)	36 M 4211		16		1
plastic cups (9 oz.)	18 M 3676	14			3
plastic play putty	15 M 8756	5			1 package
plastic wrap	15 M 9858	11, 23			1 roll
post, wooden				5	1
poster board	15 M 9856			1	1 sheet
white (15 cm × 25 cm)			7		1 piece
protractor	15 M 4067	7	29	10	1
rain gauge or coffee can	23 M 1415			4, 8	1
reference book (with information on the geologic eras)			18		1
ring stands	15 M 0660				
with rings	15 M 0707	26	24		2; 1
with rings	15 M 0709		21		1
with clamps	15 M 0698		4, 20		4, 1
rock(s)					
chips (silicate)	250 M 8108	12			500 g
samples	*		10		5
small	45 M 1986	15	16		1/4 cup; 5–6

*47 M 1037, 47 M 2222, 47 M 3307, 47 M 3512, 47 M 3637, 46 M 3822, 47 M 4672, 47 M 4662, 47 M 4807, 47 M 5742, 47 M 6442, 47 M 7057, 47 M 7262, 47 M 7402, 47 M 7607

Materials	WARD'S Catalog No.	Small-Scale	In-Depth	Long-Range	Quantities per student or group
rolling pin	15 M 0082	4, 15			1
rope	15 M 3991		22		2.5 m
rubber band	15 M 9824	6, 11, 14, 23, 24	27		1
rubber stopper	15 M 8468		20		1
ruler, metric	15 M 4650	1, 3, 4, 6, 7, 8, 13, 14, 16, 18, 19, 21, 22, 23, 25, 29, 30	2, 3, 4, 5, 6, 7, 13, 15, 16, 20, 21, 27, 28, 30	8, 9	1
safety goggles	15 M 3046	5, 16	8, 10, 11, 12, 21, 23, 29		1 pair per student
safety pin	14 M 0200	28	28		1

Materials	WARD'S Catalog No.	Small-Scale	In-Depth	Long-Range	Quantities per student or group
salt (table)	37 M 5482	21	21		1/8 cup; 2 tsp.
sand		6	15		15–20 lb.; 3–5 lb.
		13	16		3–4 lb.; 5–10 lb.
		14	20		1 cup; 1/2 lb.
coarse	45 M 1984	15	13		2 lb.; 2 lb.
fine	45 M 1983	16			5–10 lb.
quartz	45 M 1983		9		grains
sandpaper	15 M 3010			2	1 sheet
saucepan	14 M 6997	10, 11			1
saucer	18 M 7100	14			1
scissors	14 M 0525	4, 17, 21, 27, 28, 29	3, 5, 7, 21, 28, 29	1	1 pair
Secchi disc	21 M 0110			2	1
sheet metal, thin (2 cm × 8 cm)			28		1
shoe box			29		1
with lid		6, 28	28		1
sieves	13 M 0259, 13 M 0255, 13 M 0283		20		3 (4 mm, 2 mm, 0.5 mm)
sink (with faucet)		13			1
slide box	30 M 5912			5	1
soil	45 M 1990	14			1 cup
potting	20 M 8306		26		1/4 lb.
A Horizon	48 M 4220		12		1–2 g
B/C Horizon	48 M 4221		12, 20		1–2 g
silty (unsorted)	45 M 1982		13		2 lb.
spoon					
mixing	15 M 4330	7, 10			1
plastic	15 M 9800		17		1
spring, flexible/toy	16 M 0513	20			1
squeeze bottle, plastic (16 mL)	18 M 1551		7		1
steel file	14 M 0625	9	9		1
stirring rod	18 M 2010		8		1
stopwatch	15 M 0512	20	20		1
strainer	15 M 9834	12			1
streak plate	12 M 0290	9	9		1
stream table	36 M 4211		16		1
string	15 M 9863	1	24		30 cm
heavy		20			30 cm
(or fishing line)				3	7 m

Materials	WARD'S Catalog No.	Small-Scale	In-Depth	Long-Range	Quantities per student or group
sulfuric acid, dilute (3 M)	37 M 8621		11		3–4 mL
syringe, disposable (60 cc)	14 M 1620		23		1

tape
  adhesive
  masking
	
	15 M 9828
		
		3, 11, 20, 22
		23, 27, 28
			
			20
			2, 5, 8, 16,
			26, 27,
			28, 29
				
				4, 5
					1 roll

transparent/cellophane plastic	15 M 1959 15 M 5020	2		 2	 2 rolls (different colors)
teaspoon	15 M 9800	22	20, 21		1

test tubes
  13 mm × 100 mm
  20 mm × 150 mm
	
	17 M 0610
	17 M 0650
		
		10
			
			8, 11, 12
					
					2, 2, 9
					3

test-tube clamp/holder (tongs or tweezers)	15 M 0841	10	8, 11		1
test-tube rack	18 M 4231		11, 12		1
thermometers (°C) classroom	 15 M 1430	 11, 21, 24, 25, 26	 24		 1, 1, 1, 1, 3; 2
laboratory	15 M 1416		4, 21, 23, 26, 28		3, 1, 1, 2, 1
outdoor	23 M 1113			6, 8	1, 1
thread	15 M 9848	8, 27		10	1 spool; 15 cm
thread spools (small)		14			3
timer	15 M 1493		13		1
tongue depressor, wooden	14 M 0103			1	1
toothpicks, round	15 M 9840		29		6
tweezers	14 M 0515		17		1
wax pencil		16			1
weather maps				7	5 consecutive days × 2

wire
  connecting (30 cm long)
  insulated
	16 M 0549		
			8
			27
					
					2 (60 cm)
					35 cm

wire gauze	15 M 0674		21		1 square

wood, blocks
  2.5 cm × 2.5 cm × 6 cm
  5 cm × 5 cm × 5 cm
  small
  large
	16 M 0606		
			5
			15
			
			16
				
				4
					
					2
					1
					1
					1

wood, soft-wood board
  30 cm × 20 cm
  9 in. × 12 in.
	250 M 0809	
		15
			
			2
					
					1
					1

wood splint	14 M 0105		8		1